IUPAB Biophysics Series
sponsored by
The International Union of Pure and Applied Biophysics

An introduction to the physical properties of
large molecules in solution

Editors

Franklin Hutchinson
Yale University
Watson Fuller
University of Keele
Lorin J. Mullins
University of Maryland

An introduction to the physical properties of large molecules in solution

E. G. RICHARDS

With an additional chapter
The scattering of radiation by macromolecules
S. D. DOVER

Department of Biophysics, King's College
University of London

CAMBRIDGE UNIVERSITY PRESS

CAMBRIDGE

LONDON   NEW YORK   NEW ROCHELLE

MELBOURNE   SYDNEY

Published by the Press Syndicate of the University of Cambridge
The Pitt Building, Trumpington Street, Cambridge CB2 1RP
32 East 57th Street, New York, NY 10022, USA
296 Beaconsfield Parade, Middle Park, Melbourne 3206, Australia

First published 1980

Printed in the United States of America
Typeset by Automated Composition Service Inc., Lancaster, PA
Printed and bound by Vail-Ballou Press, Inc., Binghamton, NY

British Library cataloguing in publication data

Richards, Edward Graham
An introduction to the physical properties
of large molecules in solution. – (International
Union of Pure and Applied Biophysics. IUPAB
biophysics series; 3).
1. Polymers and polymerization
2. Solution (Chemistry)
I. Title   II. Dover, S D   III. Series
547'.84   QD381   79-41583

ISBN 0 521 23110 8 hard covers
ISBN 0 521 29817 2 paperback

CONTENTS

FOREWORD

The origins of this series were a number of discussions in the Education Committee and in the Council of the International Union of Pure and Applied Biophysics (IUPAB). The subject of the discussions was the writing of a textbook in biophysics; the driving force behind the talks was Professor Aharon Katchalsky, first while he was president of the Union, and later as the honorary vice-president.

As discussions progressed, the concept of a unified text was gradually replaced by that of a series of short inexpensive volumes, each devoted to a single topic. It was felt that this format would be more flexible and more suitable in the light of the rapid advances in many areas of biophysics at present. Instructors can use the volumes in various combinations according to the needs of their courses; new volumes can be issued as new fields become important and as current texts become obsolete.

The International Union of Pure and Applied Biophysics was motivated to participate in the publication of such a series for two reasons. First, the Union is in a position to give advice on the need for texts in various areas. Second, and even more important, it can help in the search for authors who have both the specific scientific background and the breadth of vision needed to organize the knowledge in their fields in a useful and lasting way.

The texts are designed for students in the last years of the standard university curriculum and for Ph.D. and M.D. candidates taking advanced courses. They should also provide a suitable introduction for someone about to begin research in a particular field of biophysics. The Union is pleased to collaborate with the Cambridge University Press in making these texts available to students and scientists throughout the world.

Franklin Hutchinson, *Yale University*
Watson Fuller, *University of Keele*
Lorin Mullins, *University of Maryland*
Editors

PREFACE

This book has its origins in a course of lectures delivered to third-year undergraduates at King's College. The course is attended by physicists, chemists and biologists as well as a sprinkling of postgraduate students. It is to such an audience that the present work is addressed.

It is, therefore, an introduction to various aspects of the nature and behaviour of large molecules in solution for the non-specialist. It is theoretical in tone and attempts to describe the nature of large molecules in general terms and to relate their behaviour to well-known and elementary principles of classical physics and chemistry. It is not concerned with particular properties of macromolecules, which depend on their individual chemical structure. Thus there is no discussion of their spectroscopic properties nor of multiple binding equilibria. Almost all of what I have written can be understood in terms of classical rather than quantum physics. I have made little attempt to review the experimentally observed facts relating to real macromolecules, but have dwelt at length on models and theories which describe in general terms their properties.

With such a non-specialist readership in mind, I have thought it wise to assume a minimum background knowledge of mathematics, physics and chemistry. I have emphasised the physical principles involved rather than mathematical detail or rigour, but, to enable the serious student to reconstruct for himself the derivation of the various results described, I have provided frequent signposts and tried to make clear the assumptions and approximations involved.

The mathematical knowledge required is no more than can be reasonably assumed to be in the possession of any third-year science student: elementary algebra, trigonometry, calculus (including some knowledge of partial differentiation) and vectors. Differential equations occur, but I merely quote their solutions. A knowledge of complex numbers is required for chapter 7, but I have provided an appendix which briefly describes their significance. A problem requiring the use of matrix algebra is considered in chapter 5, but the details of its solution is relegated to a separate appendix for the benefit of students acquainted with this branch of mathematics. Undoubtedly the text would be improved in conciseness, generality and rigour if matrix algebra and the vector calculus were to be extensively deployed, but I have thought this not to be appropriate in an elementary textbook.

I have also assumed that the reader is familiar with elementary physics and chemistry including the notions of chemical structure, the covalent bond, energy, entropy and chemical and thermodynamic equilibrium. Similarly I have assumed an elementary knowledge of mechanics, the wave nature of electromagnetic radiation, electrostatics and the Boltzmann distribution. I have attempted to present arguments from these first principles in what I hope are the simplest forms.

Although I have placed little emphasis on experimental results, I have attempted in general terms to indicate the extent to which the theories propounded are in agreement with observation and I have also described the principles behind several experimental techniques for investigating large molecules. I have, however, given few details of how such experiments are conducted or of the equipment employed.

The student working through this book (or selected parts of it) is advised to reconstruct the mathematical derivations alluded to for himself. He is also advised to work through the problems given at the end of each chapter. He should note that problems marked with an asterisk are concerned with the derivation of results rather than their application. If he requires further information on any topic he should consult the annotated list of books and review articles given at the end of the book.

As a matter of general principle, I have consistently defined physical quantities and the symbols used to denote them without prejudice to the units used in their specification. Thus the value of a quantity should be viewed as the product of a numerical value and a unit. The units employed may be chosen at will but I have used SI units or their approved multiples or sub-multiples. This principle is at variance with conventional usage in several areas, but, where confusion might arise, I have indicated so in the text.

I have attempted, wherever possible to abide by the recommendations of the Royal Society (*Quantities, Units and Symbols*, The Royal Society, London, 1971) concerning the symbols used to denote various quantities. However, the range of fields discussed in this book has made it impossible to employ each symbol in a unique manner without the use of esoteric alphabets. It is my impression that the use of symbols for which the reader has not a ready name to mind renders a text that much more difficult to follow. For this reason I have avoided the practice. To obviate any confusion that might arise I have provided a glossary of symbols that are used in a consistent manner throughout the book and have attempted to define each other symbol as it is introduced.

Finally I am grateful to Mrs Blair for typing the manuscript and to Professor F. Hutchinson and Dr R. F. Woodbridge for reading the manuscript and providing useful comments and criticism. Needless to say any errors or confusions that remain are entirely my own responsibility.

London, March 1980   *E. G. Richards*

GLOSSARY OF SYMBOLS

The physical quantities listed below are consistently denoted by the accompanying symbols. It should be noted, however, that several of the symbols are also used with other meanings, but care has been taken to ensure that these alternative usages are defined as they are introduced. The reader is also directed to table 3.1.

A	Area
a	Persistence length
a	Semi-axis of revolution of ellipsoid
a	Amplitude of wave
a_i	Activity of component i
B	Second virial coefficient
b	Semi-equatorial axis of ellipsoid
C	Third virial coefficient
C	Torque
C_σ	Characteristic ratio of polymer of σ links
C_M	Constant in expression for chain expansion
C_I	Constant in expression for expansion of polyelectrolyte
c	Velocity of light
c_i	Mole concentration of component or species i
c_c	Concentration of crosslinks
D_i	Diffusion coefficient of component i
d	Diameter of rod
E	Magnitude of electric field
E	Electric field vector
E_0	Amplitude of wave
F	Tension
F	Faraday constant
F_i	Force acting per unit mass on species i

$F(h)$	Atomic scattering factor
f	Translational frictional coefficient
f_r	Translational frictional ratio
f_0	Translational frictional coefficient of sphere with same volume as particle
f_b	Frictional coefficient of bead
f_i	Activity coefficient of species i (mole fraction scale)
G	Gibbs free energy
$G^1(\tau)$	Field autocorrelation function
$G^2(\tau)$	Intensity correlation function
g	Shear rate
g_i	Mass concentration gradient of component i
H	Enthalpy
I	Ionic strength
J_i	Flow or flux of component i
$J(\phi)$	Rotational flux
K_j	Equilibrium constant of jth reaction
K	Optical constant in light scattering
k	Boltzmann constant
k	Huggins constant
k	Circular wavenumber
L	Contour length
L	Length of rod
L_{ij}	Phenomenological coefficient
l_i	Vector representing bond i
l_i	Length of bond i
l_{av}	Average bond length
M_i	Molar mass of component i
M_0	Molar mass of monomer unit
m	Mass
m_i	Mass of component i present
m_e	Mass of electron
N_i	Number of molecules of i present
N_A	Avogadro's constant
n	Refractive index
n	Unit vector
P	Pressure
p	Dipole moment vector
p	Axial ratio b/a of ellipsoid of revolution

Q	Electrostatic charge
Q	Partition function: sum of weights of configurations
q	Charge on proton
R	Gas constant
R	Radius of sphere
R_d	Radius of domain
R_θ	Rayleigh ratio
r	End-to-end distance of polymer chain
$\boldsymbol{r}$	Vector drawn between ends of chain
S	Entropy
s	Distance from centre of mass
s_i	Sedimentation coefficient of component i
T	Temperature
t	Time
U	Internal energy
u	Excluded volume
u_i	Velocity of component i
V	Volume of system
V_d	Molar volume of domain
$V(x)$	Potential energy as function of molecular parameter x
v	Volume element
W_i, w_i	Weight fraction of component i
$W(x, y, z)$	Distribution function of end-to-end vector
w_i	Weighting factor of element i
X_σ	Mole fraction of component of degree of polymerisation σ
X_i	Thermodynamic force acting on unit mass of component i
x	Number of helical units
x_i	Mole fraction of component i
y	Number of helical regions
Z	Charge number of macroion
z	Lattice coordination number
z_i	Charge number of ion i
α	Polarisability
α	Expansion factor of polymer chain
$\alpha(x)$	Deformation ratio in x direction
β	Effective bond length

Γ	Orientation dependent factor in expression for intermolecular force
γ	Exponent of molecular weight
ϵ_0	Permittivity of free space
ϵ_r	Dielectric constant (relative permittivity)
ζ	Rotational frictional coefficient
ζ_0	Rotational frictional coefficient of sphere with same volume as particle
ζ_r	Rotational frictional ratio
ζ	Zeta potential
η	Viscosity
η_{sp}	Specific viscosity
$[\eta]$	Intrinsic viscosity
Θ	Quadrupole strength
Θ	Theta temperature
Θ	Rotational diffusion coefficient
θ	Bond angle between bonds i and bond $i-1$
θ	Helix content
κ	Debye–Hückel parameter
λ	Wavelength
λ_i	Absolute activity of component i
μ_i	Chemical potential of component i
ν	Frequency
ν	Factor in expression for intrinsic viscosity
ν_i	Electrophoretic mobility of component i
ξ	Charge condensation parameter
Π	Osmotic pressure difference
ρ	Density
ρ_i	Mass concentration of component i
ρ_0	Density of pure solvent
$\rho(s)$	Segment density distribution function
σ	Degree of polymerisation or number of segments in chain
σ	Helix initiation constant
τ	Rotational relaxation time
τ	Correlation time
Φ	Electrostatic potential
ϕ	Dihedral angle

ϕ Phase angle

$\phi(r)$ Radial distribution function of end-to-end distance

χ_i Energy factor in theory of interaction of polymer segments with solvent

Ψ Interaction parameter in theory of polymer free energy

ψ_{ij} Angle between bonds i and j

Ω Number of configurations of system

ω Circular frequency

ω Angular velocity

Some physical constants

Speed of light in a vacuum	c	2.997925×10^8 m/s
Permittivity of a vacuum	ϵ_0	$8.8541853 \times 10^{-12}$ s^{4}A^2/m^{3}kg (F/m)
Electric charge on a proton	q	$1.6021892 \times 10^{-19}$ A s
Avogadro's constant	N_A	6.022169×10^{23}/mol
Mass of electron	m_e	9.10956×10^{-31} kg
Faraday constant	F	9.64867×10^4 A s/mol
Gas constant	R	8.31434 m^{2}kg/s^{2}K mol (J/K mol)
Boltzmann constant	K	1.38062×10^{-23} m^{2}kg/s^{2}K

Properties of water at 25°C (298 K)

Density	ρ	99.7044 kg/m^3
Viscosity	η	8.937×10^{-3}/kg m s
Dielectric constant	ϵ_r	78.54
Refractive index	n	1.3329 ($\lambda = 589.3$ nm)

1 The nature of large molecules

A solution of large molecules is composed of a solvent and macromolecular solute. The solvent may be a non-aqueous solvent, such as cyclohexane, or a mixture of such solvents or it may consist mainly of water with the possible addition of salts or buffer components required to maintain a constant pH.

The macromolecular solute comprises large molecules, which we shall refer to as macromolecules, polymer molecules or particles. These are large in dimension and in molecular weight (we shall use 'molecular weight' rather than the new synonym relative molar mass) compared with the molecules of the solvent, but not so large that they sediment from the solution under the action of gravity.

The molecules of a polymer consist of a large number of smaller units, the monomer units, joined by covalent bonds. As such they are generally highly flexible objects which resemble coils of some fibre. In some circumstances, certain types of polymer may adopt a more rigid and rod-like shape; other sorts may roll up into a compact structure resembling a sphere. Particles resembling compact spheres may also be formed by the aggregation of small molecules which are then held together by secondary forces. Such aggregates are often referred to as colloidal particles. This theme of coils and rods and spheres will be found to run through this book.

In this chapter we shall be concerned to identify and to classify some of the sorts of large molecule whose physical properties are examined in later chapters.

1.1 Polymers

The simplest type of polymer molecule consists of a linear string of monomer units joined like the beads in a necklace. The monomers may be as simple as a methylene group ($-CH_2-$) in polymethylene, or of considerable complexity as are the nucleotide units (or residues) in the naturally occurring nucleic acids. The nature and chemical structure of the monomer units of a representative collection of linear polymers is illustrated in table 1.1. The number of monomer units in the chain is generally of the order of 10^2-10^4,

Table 1.1. *The chemical structure of representative linear polymers*

Name	Structure of monomer unit	Common name
(a) Synthetic polymers which do not occur in nature		
Polymethylene	$-CH_2-$	Polythene
Polytetrafluoroethylene	$-CF_2-CF_2-$	PTFE, Teflon
Polyhexamethyleneadipamide	$-NH(CH_2)_6NHCO(CH_2)_4CO-$	Nylon 66
Polyacrylamide	$-CH_2-CH-$ $\quad\quad\mid$ $\quad\quad CONH_2$	
Polystyrene	$-CH_2-CH-$ with phenyl ring	
Polymethylmethacrylate	$-CH_2-CH-$ $\quad\quad\mid$ $\quad\quad COCH_3$	Perspex, lucite
Polydimethyl siloxane	$\quad\quad CH_3$ $\quad\quad\mid$ $-O-Si-$ $\quad\quad\mid$ $\quad\quad CH_3$	Silicone
Polyphosphoric acid	$\quad\quad OH$ $\quad\quad\mid$ $-O-P-$ $\quad\quad\parallel$ $\quad\quad O$	Calgon (Na salt)
Polyvinylamine	$-CH_2-CH-$ $\quad\quad\mid$ $\quad\quad NH_2$	
Polyacrylic acid	$-CH_2-CH-$ $\quad\quad\mid$ $\quad\quad COOH$	
Polyvinyl sulphonic acid	$-CH_2-CH-$ $\quad\quad\mid$ $\quad\quad SO_3H$	
(b) Polymers which occur in living organisms		
Polypeptides	$-NH-CH-CO-$ $\quad\quad\mid$ $\quad\quad R$	Proteins
Polynucleotides	$B\quad O\quad CH_2-O-PO_2-$ ring with OH, *HO	RNA, DNA (DNA lacks the OH group*)

Side chains ($-R$) found in naturally occurring proteins and the common names of the amino acids from which the monomer units are derived:

$-H$	Glycine (gly)
$-CH_3$	Alanine (ala)
$-CH(CH_3)_2$	Valine (val)
$-CH_2CH(CH_3)_2$	Leucine (leu)
$-CH(CH_3)(CH_2CH_3)$	Isoleucine (ilu)
$-(CH_2)_2-S-CH_3$	Methionine (met)
$-CH_2-Ph$	Phenylalanine (phe)

Table 1.1. *(cont.)*

Name	Structure of monomer unit	Common name

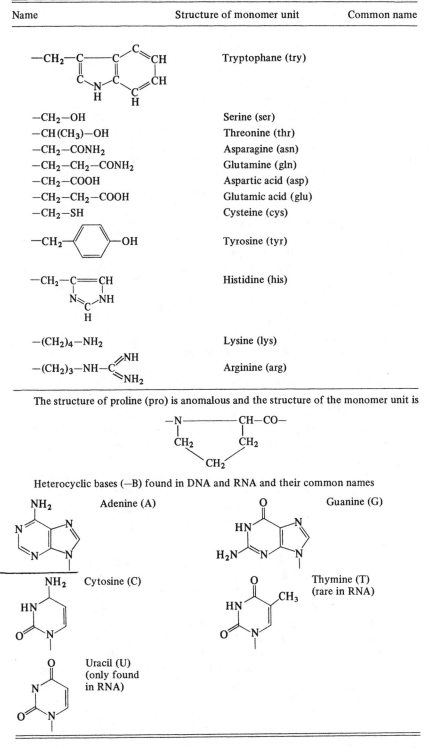

		Tryptophane (try)
$-CH_2-OH$		Serine (ser)
$-CH(CH_3)-OH$		Threonine (thr)
$-CH_2-CONH_2$		Asparagine (asn)
$-CH_2-CH_2-CONH_2$		Glutamine (gln)
$-CH_2-COOH$		Aspartic acid (asp)
$-CH_2-CH_2-COOH$		Glutamic acid (glu)
$-CH_2-SH$		Cysteine (cys)
		Tyrosine (tyr)
		Histidine (his)
$-(CH_2)_4-NH_2$		Lysine (lys)
		Arginine (arg)

The structure of proline (pro) is anomalous and the structure of the monomer unit is

Heterocyclic bases (−B) found in DNA and RNA and their common names

Adenine (A)

Guanine (G)

Cytosine (C)

Thymine (T)
(rare in RNA)

Uracil (U)
(only found
in RNA)

though polymers with a considerably greater number are not unknown. This number, σ, is sometimes referred to as the degree of polymerisation.

It is worth noting that similar structures with very much smaller degrees of polymerisation are referred to as *oligomers*. The simplest of these is the *dimer*, followed by the *trimer*, *tetramer* etc. which contain two, three, four etc. monomers respectively.

Polymers in which the monomer units are all the same are called *homopolymers;* synthetic polymers used in the plastics and textile industries are mostly (though not invariably) of this kind. If two or more different monomer units are present we have a *copolymer* or *heteropolymer*. In the naturally occurring proteins, any one of the monomer residues may be derived from any one of twenty different amino acids as indicated in table 1.1(b). Similarly, in the polynucleotides, any one of four (and in some cases more) heterocyclic bases may occur. In living cells there are elaborate mechanisms for ensuring that the precise sequence of amino-acid residues in proteins or of nucleotide residues in DNA and RNA is constant. This precise sequence of amino-acid residues or nucleotide residues constitutes the *primary structure* of the protein or polynucleotide. Synthetic copolymers may also be prepared in which the sequence of the different monomer residues is essentially random.

A linear polymer may have a direction associated with it. That is to say that if you start at one end and move towards the other, the molecular landmarks you will encounter will be in a different order than if you start at the other end. Polypeptides and polynucleotides have this property, but polymethylene has not. Another way of recognising a chain with direction is to stretch it out and then reverse it by a rotation of $180°$; if the reversed and non-reversed chains are identical, the chain has no direction.

We have seen that a linear polymer consists of a chain of monomer units joined by covalent bonds. In such polymers we may generally distinguish a sequence of atoms joined by covalent bonds which runs from one end of the chain to the other. This sequence defines the *backbone* of the polymer. In polymethylene, this is a sequence of carbon atoms, but in a polypeptide it is a sequence $[-N-C-C-]_n$. According to the detailed structure of the monomer unit, various other atoms or chemical groups are joined to the backbone atoms.

An interesting and subtle way in which the monomer units along the chain may differ is exhibited by certain polymers including polystyrene. In these one or more carbon atoms in the backbone of each monomer are joined to four different groups. Such carbon atoms constitute a centre of asymmetry. If the backbone of polystyrene is stretched out to form a zig-zag, with all the carbon atoms in the same plane, each phenyl group may lie either above or below this plane as illustrated in figure 1.1. If they are all on the same side, the polymer is *isotactic*. It might be thought that there would be two different isotactic polystyrenes: one with the phenyl groups above the plane and

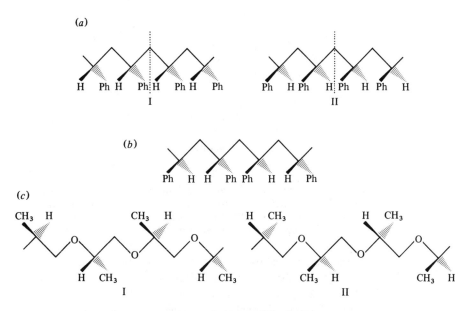

Figure 1.1. Examples of stereoregular polymers: (*a*) isotactic polystyrene, I and II are in fact identical, as may be seen by rotating II about the axis; (*b*) syndiotactic polystyrene; (*c*) isotactic poly(propylene oxide), I and II are mirror images.

the other with them below the plane. However, if we turn one of these forms through 180°, we find it to be identical to the other. If the phenyl groups are randomly disposed above and below the plane the polymer is *atactic*. A further possibility arises if the groups are alternately above and below, in which case the polymer is *syndiotactic*. If the number of backbone atoms in each monomer is odd, the two forms of the isotactic polymer may be different; such a possibility is illustrated in figure 1.1(*c*) for isotactic poly(propylene oxide). Note that the isotactic polymer is defined in this case to be the form in which each asymmetric carbon atom is in the same stereochemical form. This is the more general and more usual definition. Polymers exhibiting these characteristics are termed *stereoregular* polymers.

Linear polymers are but the most simple type of polymer. In them each monomer unit is bifunctional; that is to say each unit (except the two at the ends) is joined to two others. Other polymers may contain trifunctional or tetrafunctional units. A trifunctional unit in a polymer leads to a branch point where the chain may proceed in either of two directions. The introduction of only a small proportion of trifunctional units among the monomers can lead to a complicated ramifying structure. The naturally occurring polysaccharide, cellulose, is an example of such a structure. According to the

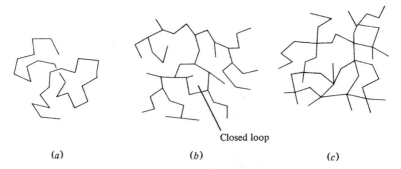

Closed loop

(a)   (b)   (c)

Figure 1.2. Two-dimensional representations of: (a) Linear polymer containing only bifunctional units; (b) branched-chain polymer containing both bifunctional and trifunctional units; (c) cross-linked network with bifunctional and tetrafunctional units.

mode in which the polymer is assembled from its monomer units, closed loops of polymer chain can also result from trifunctional units being present as illustrated in figure 1.2.

The introduction of tetrafunctional units among the monomers leads to the possibility of cross linking the polymer chains. It is thus possible for each polymer chain (defined as a sequence of bifunctional units) in a solution to be linked to another to form a single giant molecule pervading the entire volume of the solution. If there are n such chains, $n - 1$ cross links are required to link them all together. Any additional cross links will then lead to the formation of closed loops. With a sufficient profusion of cross links, the giant molecule becomes a complicated three-dimensional net and is an example of a *gel*. Cured rubber and cross-linked polyacrylamide gels are examples of such structures in which the cross links result from the presence of covalently constructed tetrafunctional units. In other gels, such as gelatine or agar, the cross links are the result of secondary interactions between monomer units which dissociate when the gel is heated.

A further class of polymer is provided by the *polyelectrolytes*. Polyelectrolyte molecules (linear, branched or cross linked) carry an electric charge. This electric charge arises from the ionisation of appropriate groups on the monomer units. If the charge so produced is negative the ionised polymer becomes a *polyanion*, while if it is positive it becomes a *polycation*. If the polymer is a heteropolymer, and some of the monomer units can ionise to yield a positive charge and others to yield a negative charge, the polymer is a *polyampholyte*; the naturally occurring proteins are good examples of polyampholytes.

The monomer units of some polyelectrolytes behave like strong electrolytes, so that in an ionising solvent the polyion always carries its maximum charge. Such are poly-*N*-butyl-4-pyridinium bromide and polyvinyl sulphonic acid. In others, the monomer units resemble weak electrolytes, and the charge carried by the polyion depends on the degree of ionisation and hence on the pH of the solution. Polyvinylamine and polyacrylic acid are examples. When a polyelectrolyte ionises, the low-molecular-weight ions that dissociate from it are referred to as the counterions. It is clear that the total charge carried by all the polyions in the solution must equal the total charge carried by the counterions.

1.2 The size and shape of macromolecules

With polymers, linear, branched or cross linked, there is the possibility of rotation about any single covalent bond in the backbone. This means that they are often highly flexible objects and in solution they must be considered to be in continual motion, writhing and twisting and changing their shape. A polymer molecule that behaves like this is said to be a *random coil*. Such a molecule cannot be said to have a well-defined size or shape.

With linear polymers it is possible to define the *contour* length, L, as the distance between the two ends when the backbone is stretched out to its fullest extent. This parameter however gives a misleading impression of the end-to-end distance if the polymer is a random coil. In this case one may define an average end-to-end distance as the square root of the mean square end-to-end distance, $\langle r^2 \rangle$. This may be several orders of magnitude smaller than the contour length.

Under some circumstances, interactions between neighbouring parts of the polymer chain may severely restrict freedom of rotation about the bonds in the backbone so that the molecule is much more rigid and may resemble a rod. In such circumstances the polymer backbone is arranged in a helical configuration, and the end-to-end distance of the rod is proportional to the contour length. Polypeptides can thus adopt a helical rod-like conformation (the α-helix) and two polynucleotide chains can associate to form a double helical rod-like complex. Structures of this sort in proteins and polynucleotides are often said to constitute their *secondary structure*.

It is also possible for interactions between more distantly separated portions of the chain to cause it to coil up into a compact approximately spherical conformation with a reasonably well-defined radius. This radius will be (very roughly) proportional to the cube root of the number of monomer units in the chain. Certain proteins, the enzymes in particular, exhibit this ability to form compact globular structures. The pattern of folding manifest by the globular proteins defines their *tertiary structure*.

An important characteristic of polymers which form a secondary or tertiary structure is their capacity to denature under the influence of heat or a change in the solvent and revert to a random coil.

A detailed exploration of the manner in which the dimensions of random coils depend on the structure of the polymer will be presented in chapter 4, the details of the denaturation of helical and globular structures will be discussed in chapter 5.

1.3   Colloidal particles

Bearing a resemblance to the globular proteins are colloidal particles. The term colloid is usually applied to a two-phase system consisting of the *dispersed phase* and the *dispersive medium*. The material in the dispersed phase is broken up into a large number of particles each of a size qualifying them to be called large molecules and each surrounded by the dispersive medium which plays the role of solvent. A phase boundary or surface separates each particle from the dispersive medium, and many properties of the colloid (such as its stability) are dominated by the properties of this surface on account of the fact that the ratio of the surface area to the volume of the colloidal particles is large.

Each of the two phases may be a liquid, a solid or a gas, giving eight different types of colloidal system (the ninth, in which both phases are gases, obviously does not exist because all gases are miscible in all proportions). These are named, and examples are given, in table 1.2. We shall only be concerned with sols, in which the dispersive medium is a liquid and the dispersed phase is solid or liquid. Solutions of polymers, linear or branched, were sometimes referred to as colloidal systems, but there is little to be gained by calling a large molecule a separate phase, and the terminology is falling into disuse.

The colloidal particles of the dispersed phase in a sol consist of an aggregate of small molecules or atoms held together by secondary forces. These small molecules may also exist in the dispersive medium in true solution and be in equilibrium with the aggregates.

There are many different types of sol particle, and these vary greatly in their nature. They may be metallic, as in colloidal gold (purple of Cassius), or crystalline, as in a colloidal suspension of silver chloride. A particularly interesting type of colloidal particle arises from the aggregation of lipid and detergent molecules. Molecules of these compounds consist of a hydrophilic 'head', which has an affinity for water, and a hydrophobic 'tail', which has not. Such molecules tend to aggregate when dissolved in water in such a way that the hydrophobic tails are clustered together out of contact with the water and such that the hydrophilic heads are exposed to it. Some of the sorts of structure that can arise from aggregations of this kind are illustrated in figure 1.3.

The individual molecules that aggregate to form the colloidal particles can themselves be macromolecular. Thus several protein molecules (of one or several kinds) may aggregate in a specific manner to form a larger entity. The manner of arrangement of the subunits in such multisubunit proteins constitutes their *quaternary structure*. In a similar manner molecules of protein may aggregate with polynucleotides to form virus particles and ribosomal

Table 1.2. *The classification of colloidal systems*

The dispersed phase	The dispersive medium		
	Gas	Liquid	Solid
Gas	Gases mix in all proportions	Foam (whipped cream, beaten egg white)	Solid foam (sponge rubber)
Liquid	Mist (steam from a kettle, clouds)	Emulsion (cream, mayonnaise)	Solid emulsion (butter)
Solid	Smoke	Sol (see text for examples)	Solid suspension (various minerals)

Note that the mechanical properties of the colloidal system are dominated by those of the dispersive medium.

particles. Indeed the various supramolecular structures and organelles found in living cells are, in the final analysis, aggregates of proteins, nucleic acids and lipids, arranged in ever more complicated patterns.

Colloidal solutions may be classified into two categories: the *lyophilic* colloids and the *lyophobic* colloids. The surfaces of the particles of lyophilic colloids interact strongly with the solvent, and it is this interaction which is mainly responsible for preventing their aggregation into larger particles and maintaining their stability in solution. The particles of lyophobic colloids, on the other hand, have no such affinity for the solvent and, in the absence of compensating features, would tend to flocculate and precipitate out of solution. These compensating features are provided by electric charges on their surfaces. This gives rise to electrostatic repulsions between the particles, which prevents them from aggregating. This electric charge may originate from the ionisation of the small molecules of which they are composed or from the absorption of ions provided by electrolytes present in the solution. Gold sols, silver halide and other inorganic sols tend to be lyophobic. Proteins (in so far as it is profitable to consider them as colloids) and micelles tend to be lyophilic colloids.

1.4   Molecular-weight distributions

Some polymer molecules, proteins and some nucleic acids in particular, have well-defined molecular weights and well-defined degrees of polymerisation. Others, including most synthetic polymers and colloidal particles, have molecular weights that span a range. If this range is narrow they are termed *paucidisperse*, while if it is broad they are termed *polydisperse*. If the molecular weights of all the macromolecules present are the same they are termed *monodisperse*.

Dodecyl sulphate:

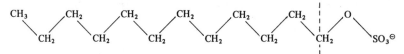

CH₃ — CH₂ — CH₂ — CH₂ — CH₂ — CH₂ — CH₂ — O — SO₃⁻

Hydrophobic 'tail'   Polar 'head group'

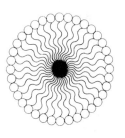

Micelle in polar
solvent

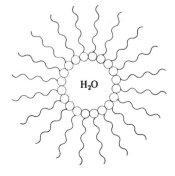

H₂O

Micelle in non-polar
solvent

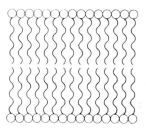

Bilayer structure
formed by lipids
in living cells

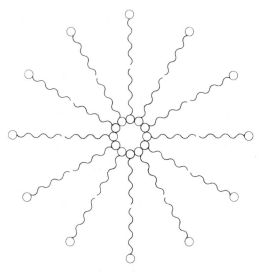

Spherical vesicle formed
from closed bilayer

Figure 1.3. Example of a polar molecule which forms micelles, and examples of the sorts of structures that such molecules can form.

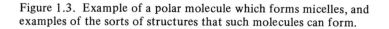

Let us consider explicitly linear homopolymers but bear in mind that the concepts of molecular-weight distributions and averages that we shall discuss apply equally well to branched and heteropolymers and to colloidal particles. In the case of homopolymers the molecular weight, M_σ, of a particular molecule is proportional to its degree of polymerisation, σ, so that $M_\sigma = \sigma M_0$, where M_0 is the molecular weight of the monomer unit. In the case of heteropolymers, M_0 must be defined as the average molecular weight of the monomer units in the chain.

Let us suppose that in a polydisperse polymer sample the number of molecules with a degree of polymerisation σ is N_σ. The molecular weight of these molecules is M_σ, which is an integer multiple of M_0. We may plot N_σ against M_σ or σ in the form of a histogram but, if we are dealing with high degrees of polymerisation, this plot approximates to a smooth curve as illustrated in figure 1.4(*a*). Since the N_σ are large numbers it is convenient to divide each N_σ by the total number of polymer molecules present, which is $\Sigma_{\sigma=1}^{\infty} N_\sigma$. The resulting ratios, X_σ, define the mole fractions of the various species in the polymer mixture. Plots of X_σ against M_σ resemble plots of N_σ with only a change in scale, but the area under the curve is now equal to unity. Curves of this kind are called number distribution curves since they define the number (or mole fraction) of molecules with a given molecular weight. An alternative method of displaying the distribution of molecular weights is to plot the mass of each species present against M_σ. The mass of the species with molecular weight M_σ that is present in the mixture is clearly proportional to $N_\sigma M_\sigma$. It is convenient to normalise such distribution curves by dividing the mass $N_\sigma M_\sigma$ by the total mass of polymer present, $\Sigma_{\sigma=1}^{\infty} N_\sigma M_\sigma$, to obtain the weight fraction, W_σ, of each species present. Plots of W_σ against M_σ or σ define the weight distribution curve. The number and weight distribution curves for a typical condensation polymer are compared in figure 1.4(*b*) (for a discussion of condensation polymerisation, the reader should consult Tanford (1961) in the further reading section for chapter 1).

Data which would enable distribution curves to be plotted are not readily available for most polymer preparations. It is only with great diligence that it is possible to fractionate a polymer preparation into a series of paucidisperse fractions that can be used, after their molecular weights have been measured, to define the distribution curve. It is also, in favourable cases, possible to determine the weight distribution curve directly. Usually, however, all that is either expedient or possible is to measure an average molecular weight.

The simplest average molecular weight is the number average; this is defined by the relation

$$M_n = \frac{\sum N_\sigma M_\sigma}{\sum N_\sigma} = \sum X_\sigma M_\sigma \qquad (1.1)$$

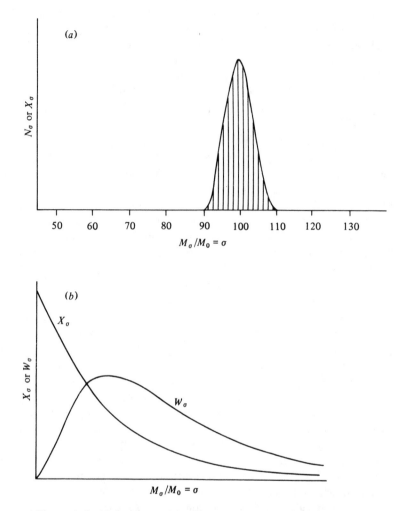

Figure 1.4. Molecular weight distributions. (a) Histogram showing number of molecules (or their mole fractions) with different molecular weights, M_σ, or degrees of polymerisation, σ. The histogram may be represented by the smooth curve enclosing it. (b) Number-distribution and weight-distribution curves for a typical condensation polymer. The mole fraction, X_σ, may be represented by $p^{\sigma-1}(1-p)$ and the weight fraction, W_σ, by $\sigma p^{\sigma-1}(1-p)^2$, where p is a constant depending on the details of the polymerization process.

In this average, each molecular weight is weighted according to the mole fraction of the corresponding species in the mixture. If, alternatively, they are weighted according to the weight fraction of the species, the weight-average molecular weight is obtained as defined by

$$M_w = \frac{\sum N_\sigma M_\sigma^2}{\sum N_\sigma M_\sigma} = \sum W_\sigma M_\sigma \tag{1.2}$$

It should be noted that the summations in equations (1.1) and (1.2) are taken over all possible values of the degree of polymerisation, σ, that is for $\sigma = 1$ to $\sigma = \infty$.

It may be noted that the formula for the weight average is derived from that of the number average by multiplying each term in both the numerator and the denominator by the appropriate value of M_σ. This process may be repeated to yield the z-average molecular weight defined by

$$M_z = \frac{\sum N_\sigma M_\sigma^3}{\sum N_\sigma M_\sigma^2} \tag{1.3}$$

Similarly, the $(z + 1)$ average, the $(z + 2)$ average etc. $(M_{z+1}, M_{z+2}$ etc.) may be defined by further repetitions. The formula for the $(z + j)$-average molecular weight is thus

$$M_{z+j} = \frac{\sum N_\sigma M_\sigma^{3+j}}{\sum N_\sigma M_\sigma^{2+j}} \tag{1.4}$$

For a monodisperse polymer all these different averages are equal; for a polydisperse polymer, on the other hand, the higher members of this series of averages are greater than the lower members because they give more weight to the higher-molecular-weight species; thus we have

$$M_n \leqslant M_w \leqslant M_z \leqslant M_{z+1} \leqslant \cdots \tag{1.5}$$

There are experimental techniques that are capable of yielding at least the lower members of this series of averages. It should also be noted that there are other experimental methods that yield average molecular weights that do not belong to this series. Thus in chapter 8 it will be shown that the intrinsic viscosity of a monodisperse polymer is proportional to M^γ, where γ is some constant which depends on the nature of the polymer. If the intrinsic viscosity of a polydisperse sample is measured, and each individual species follows this relation, a viscosity-average molecular weight is obtained which is given by

$$M_\eta = \left[\frac{\sum N_\sigma M_\sigma^{\gamma+1}}{\sum N_\sigma M_\sigma} \right]^{1/\gamma} \tag{1.6}$$

If γ is less than unity (for a random-coil polymer $\gamma = 0.5$), the viscosity average falls between M_n and M_w.

Problems

1.1 A certain protein associates into dimers and tetramers in the molar proportions of monomers : dimers : tetramers in the ratio $4:2:1$. Calculate the number-average, weight-average and z-average molecular weights of the mixture assuming that the molecular weight of the monomer is 100 000.

1.2 Derive expressions in terms of p for the number-average, weight-average and z-average degrees of polymerisation of a condensation polymer given that the mole fraction of the species with degree of polymerisation σ is given by $(1 - p)\, p^{\sigma - 1}$.

1.3 A sample of RNA is prepared by removing the proteins from a sample of purified ribosomes. The resulting solution contains equimolar amounts of '23S RNA', '16S RNA' and '5S RNA', with molecular weights 1.1×10^6, 0.55×10^6 and 4.0×10^4 respectively. Calculate the viscosity-average molecular weight of the mixture by assuming that the intrinsic viscosity of RNA is proportional to the square root of the molecular weight. Would the viscosity-average molecular weight change appreciably if the '5S RNA' were to be removed from the mixture?

2 Molecular interactions

We have seen in chapter 1 that colloidal particles are aggregates of smaller molecules. In this chapter we shall be concerned with the secondary interactions between the constituent parts of such particles. Such secondary interactions also occur between the different monomer units of polymers. If they occur between monomer units on two different polymer molecules they constitute intermolecular interactions. As such they play a role in maintaining the integrity of the quaternary structure of multisubunit proteins and particles composed of nucleic acids and proteins. If, on the other hand, they act between monomer units, both on the same polymer molecule, they are termed intramolecular interactions. These play a role in stabilising the secondary and tertiary structure of helical polypeptides, polynucleotides and the globular proteins. Secondary interactions also occur between the monomer units of a polymer and the surrounding solvent molecules, and indeed between one solvent molecule and another.

In this chapter we will be concerned with the nature of these secondary interactions with a view mainly to understanding intermolecular and intramolecular interactions between polymers and the effect of the solvent upon them.

2.1 Secondary forces

Secondary forces, as opposed to covalent-bond forces, act between all molecules and chemical groups and are primarily electrostatic in nature.

Consider, for simplicity, two atoms A and B separated by a distance, r, between their nuclei. There will be a force acting between them which may be either repulsive or attractive. This force will vary in magnitude with the distance r. It is more convenient however, rather than to consider the force explicitly, to consider the energy of interaction. This is the energy required to bring the isolated atoms from a great distance (infinity) apart to the separation r. We shall denote this energy, which varies with r, by $V(r)$. If $V(r)$ increases as the atoms are brought still closer together, the force between them is repulsive whereas, if it decreases, the force is attractive. If the energy

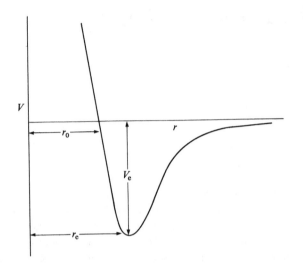

Figure 2.1. General form of the dependence of the intermolecular interaction energy, $V(r)$, on molecular separation, r. V_e is the decrease in energy when the two molecules are brought from infinite separation to the equilibrium separation, r_e, at the minimum of the curve. r_0 *is the* separation at which the energy is zero and is approximately equal to the van der Waals radius. The graph has been constructed on the basis of equation (2.2).

is at a minimum, there is no force between them, and the atoms are at their equilibrium separation. The manner in which $V(r)$ might be expected to vary with r is illustrated in figure 2.1.

The energy of interaction between two atoms at their equilibrium separation with attractive secondary forces operating is typically much less than 100 kJ/mol. This figure may be contrasted with the energy when a covalent bond is formed between the two atoms; this is of the order of 100–1000 kJ/mol. It should also be compared with the value of RT where T is the absolute temperature and R the gas constant. At room temperature (say 300 K) this is 2.5 kJ/mol. We thus see that our two atoms at their equilibrium separation may frequently acquire sufficient energy through molecular bombardment to fly apart. Conversely, this will rarely happen if they are covalently linked. Thus covalent bonds are stable and rarely break in this manner, whereas secondary bonds or interactions between atoms and molecules are unstable and may break and reform rapidly, perhaps at a rate of 10^{10} per second.

These secondary forces arise in several different ways, and may be categorised as multipole interactions, induction forces, dispersion forces and exclusion forces. The first three of these are essentially electrostatic and can be understood in terms of classical electrostatic theory without recourse to quantum theory. The exclusion forces are, however, essentially quantum mechanical in nature. Other forces such as magnetic forces also exist, but their energy is generally negligible compared with that of the others.

2.2 Multipole interactions

Any molecule or group of atoms consists of positively charged nuclei and the associated negatively charged electrons. The latter may be considered to be distributed about the nuclei in a manner dictated by the laws of quantum mechanics. Thus, in the vicinity of the molecule, each point in space is associated with an average charge density, ρ, so that a volume element dv contains an average charge element $\rho \, dv$.

The energy of interaction of two point charges Q_1 and Q_2 separated by a distance, r, in a vacuum is given by Coulomb's law as $Q_1 Q_2 / 4\pi\epsilon_0 r$, where ϵ_0 is the permittivity of a vacuum. It follows that the energy of interaction between a charge element $\rho_1 \, dv_1$ in a volume element dv_1 in the vicinity of molecule 1 and a charge element $\rho_2 \, dv_2$ associated with molecule 2 is given by $\rho_1 \rho_2 \, dv_1 \, dv_2 / 4\pi\epsilon_0 r_{12}$, where r_{12} is the distance between the two volume elements. The total energy of interaction between the two molecules must then be obtained by integrating this expression over the volume of both molecules.

This integration is plainly impossible without prior knowledge of the charge density functions, ρ. Nevertheless, it may be shown that any collection of charges, such as a molecule with its electrons and nuclei, is equivalent to a set of components called *multipoles*. This is true whatever the details of the actual distributions.

It then follows that the total energy of interaction between the two molecules can be expressed as the sum of the interactions between the various multipoles describing the charge distribution of one molecule with those describing the distribution of the other. In principle there is, associated with any charge distribution, an infinite set of multipoles termed *monopoles, dipoles, quadrupoles, octupoles* etc., but it turns out that only interactions involving the first three are of sufficient magnitude to be significant.

The monopole component turns out to be equal to the total net charge on the molecule. This is zero unless the molecule or group is an ion, in which case interactions involving monopoles dominate the interaction energy. Such interactions will be found in chapter 9 to be of crucial importance in determining the properties of polyelectrolytes.

Even if a molecule has no net charge, it may still have a dipole component, generally called a *dipole moment*. This will occur if the centres of gravity (or more properly the centres of charge) of the electrons and of the nuclei are not coincident. The magnitude of the dipole moment is equal to the product of the total electronic charge multiplied by the separation of the two charge centres. Since the separation of the centres of charge has both magnitude and direction it is a vector, and so too is the dipole moment. Conventionally it points from the negative to the positive centre of charge. Chemical bonds involving strongly electronegative atoms such as N, O or F tend to be associated with large dipole moments acting in the direction of the bond. On the other hand, carbon–carbon bonds and carbon–hydrogen bonds have but small dipole moments associated with them; symmetrical molecules such as N_2 or O_2 have zero dipole moment. If a molecule or bond has a large dipole moment, this will tend to dominate the multipole interaction energy in the absence of monopoles. Such molecules are termed *polar molecules*. In the absence of large dipole moments they are termed *non-polar molecules*.

The quadrupole moment is a still more complicated entity which requires six components for its specification: dipoles being vectors require three, and monopoles being scalars require one. The quadrupole moment is associated with the second moment of the charge distribution, and a neutral, non-polar and symmetrical molecule may still possess a finite quadrupole moment. Interactions involving quadrupole moments are generally small and comparable in magnitude to the effect of induction and dispersion forces. If a molecule has axial symmetry, a simplification in the specification of its quadrupole moment is possible, and only one parameter is required.

We must now consider the interactions between pairs of multipoles. There are six possible types of interactions between the three significant types of multipole as indicated in table 2.1.

The energies of interaction for the various pairs fall off with distance in different ways as also indicated by the formulae in rows $3(a)$, $4(a)$ and $5(a)$ of this table. In these formulae, Q, p and Θ represent the strengths of a monopole, a dipole and a quadrupole respectively. ϵ_0 is the permittivity of a vacuum and ϵ_r the relative permittivity of the medium (dielectric constant) in which they are immersed. For a vacuum, ϵ_r is unity. Γ is a factor which depends on the orientation of the multipoles. In the case of interactions involving dipoles and quadrupoles the energy depends on the orientation of both molecules with respect to the line joining them, and, according to these orientations, the energy may be positive or negative. For instance, in the case of dipole–dipole interactions, the energy is most negative (and the force between them attractive) when they are arranged head to tail; the energy is most positive (and the force repulsive) when they are arranged head to head or tail to tail. When arranged perpendicular to each other the energy of interaction is zero. If we consider the energy of interaction averaged over all possible orientations, we find that the result is zero. However all orientations are not equally probable and in general we would find, if we weighted the orientations appropriately, that there was a net attractive force between two dipoles.

Explicit expressions for the factors, Γ, which show in detail how the energy

Table 2.1. *Electrostatic interactions between molecules*

1		2	3	4	5
1		Monopole Q_2	Dipole p_2	Quadrupole Θ_2	Dispersion interaction
2	Field strength	$Q/4\pi\epsilon_r\epsilon_0 r^2$	$p\Gamma/4\pi\epsilon_r\epsilon_0 r^3$	$\Theta\Gamma/4\pi\epsilon_r\epsilon_0 r^4$	—
3	Monopole Q_1 1.6×10^{-19} C	a $\quad Q_1Q_2/4\pi\epsilon_r\epsilon_0 r$ b $\quad$ 463 kJ/mol c $\quad$ 56 nm	$Q_1 p_1\Gamma/4\pi\epsilon_r\epsilon_0 r^2$ 59 kJ/mol 1.46 nm	$Q\Theta_2\Gamma/4\pi\epsilon_r\epsilon_0 r^3$ 20 kJ/mol 0.60 nm	— — —
4	Dipole p_1 6.1×10^{-30} C m	a $\quad$ — b $\quad$ — c $\quad$ —	$p_1 p_2\Gamma/2\pi\epsilon_r\epsilon_0 r^3$ 15 kJ/mol 0.55 nm	$3p_1\Theta\Gamma/8\pi\epsilon_r\epsilon_0 r^4$ 3.7 kJ/mol 0.33 nm	— — —
5	Quadrupole Θ_1 6.1×10^{-40} C m^2	a $\quad$ — b $\quad$ — c $\quad$ —	— — —	$\Theta_1\Theta_2\Gamma/4\pi\epsilon_r\epsilon_0 r^5$ 0.83 kJ/mol 0.24 nm	— — —
6	Induced dipole	a $\quad Q_2^2\alpha_1/8\pi\epsilon_r\epsilon_0 r^4$ b $\quad$ 13 kJ/mol c $\quad$ 0.45 nm	$p_2^2\alpha_1\Gamma/4\pi\epsilon_r\epsilon_0 r^6$ 0.41 kJ/mol 0.22 nm	$3\Theta_2^2\alpha_1\Gamma/8\pi\epsilon_r\epsilon_0 r^8$ 0.07 kJ/mol 0.19 nm	$3\alpha_1\alpha_2 h\nu/4\pi r^6$ 3.4 kJ/mol 0.31 nm

depends on the orientation, can be derived, but these are mostly of some complexity. We content ourselves by giving in table 2.1 the expressions for the energies of interaction between the six possible pairs of multipole when they are orientated to give the maximum attractive force. It is worth noting that the form of these expressions may be easily checked by the method of dimensions if it is borne in mind that the SI unit of ϵ_0 is F/m, that of a monopole is C, that of a dipole is Cm and that of a quadrupole is Cm^2.

Also in column 1 of table 2.1 we give for illustrative purposes the magnitude of the monopole strength (net charge) on a monovalent ion and the dipole and quadrupole moments of a water molecule (H_2O). As well as this the table gives in rows 3(b), 4(b) and 5(b) the maximum energy of interaction between the various multipole components of two water molecules (or, in the case of interactions involving monopoles, the interaction with a single electronic charge) separated by a distance of 0.3 nm (approximately the intermolecular separation in liquid water); and in rows 3(c), 4(c) and 5(c) the separation when the energy of interaction is equal to RT with T = 300 K. In these calculations ϵ_r was assumed to be unity and ϵ_0 to be 8.854 × 10^{-12} F/m.

Two points should be noted from table 2.1. Firstly, the energy of interaction varies with the intermolecular separation raised to different inverse powers for the different sorts of interaction. This means that the relative magnitudes of the different sorts of interaction vary with separation, and that the interactions involving the higher multipoles are very short range interactions. Secondly, only interactions involving monopoles even approach the strength of covalent bonds.

One final point concerns the electric field in the vicinity of a molecule. This too can be considered as the sum of the fields due to various multipoles into which the charge distribution has been resolved. The field strengths due to the different multipole components also decrease with distance from the molecule in proportion to various inverse powers of the distance as indicated by the formulae in row 2 of table 2.1.

2.3   Induction forces

So far we have tacitly assumed that the charge distribution associated with a molecule is rigid. In actual fact the electric field due to one molecule distorts the charge distribution of another in its vicinity as if it were elastic. In doing so it *induces* an additional dipole component. This comes about because the nuclei and the electrons, having opposite charges, are pulled in opposite directions by the field. The effect is mutual, for the field of one of the molecules polarises the other.

If the molecule that is polarised in this manner is symmetrical, the induced dipole, p (which is a vector), is proportional to the electric field, E, which produces it, and the dipole is parallel to the field. In this case we may write

$$p = \alpha E \tag{2.1}$$

The constant of proportionality, α, is termed the polarisability of the molecule. If the polarised molecule is not symmetric, the dipole may then not be parallel to the field, and six quantities are required to specify the relation between them. In this case the polarisability is described as being a tensor. We shall, however, suppose that a single scalar polarisability is sufficient.

The induced dipole, p, is only produced by the field at the expense of the latter doing work in separating the charges. This work amounts to $\frac{1}{2}\alpha E^2$. However, once the induced dipole is set up, it interacts with the field that produced it to give a negative interaction energy $-\alpha E^2$. Thus the net energy of interaction is $-\frac{1}{2}\alpha E^2$. This implies that there is a force between the two molecules which is always attractive. Such forces are termed *induction forces*. It should be noted that the energy of interaction depends on the square of the field strength. This means that we cannot sum the energies of interaction of an induced dipole with the separate fields produced by the various multipoles in its vicinity. It is necessary to sum the various components of the field first and square the result. This complication is to some extent mitigated by the observation that, except at small intermolecular separations, a monopole field dominates a dipole field and the latter dominates that due to a quadrupole. This means that the field may be assumed to be that due to the lowest multipole present to a reasonable approximation.

Row $6(a)$ of table 2.1 gives formulae for the energy of interaction of a molecule with polarisability, α_1, with the fields due to various multipoles at a distance, r. Rows $6(b)$ and $6(c)$ also give estimates of the induced dipole interaction energies produced by fields equivalent to those produced by a single electron, a molecule with a dipole equal to that of a water molecule, and a molecule with a quadrupole moment equal to that of water. The value of the polarisability is assumed to be that of the water molecule (1.48×10^{-30} m^3), and the intermolecular separation 0.3 nm. The table also gives the separation required to produce an interaction energy equal to RT at 300 K.

It may be noted from table 2.1 that induced dipole interaction energies are but small if the polarising molecule is electrically neutral and has no dipole moment.

2.4   Dispersion forces

From what has been said above, it might be inferred that only quadrupole–quadrupole and quadrupole–induced-dipole interactions between two uncharged non-polar molecules need be considered, and, hence, since the energies of these are both small, such molecules interact with but negligible energy. This would be to ignore dispersion forces.

We have tacitly assumed so far that the charge distribution on a molecule is time independent. In actual fact it undergoes time-dependent fluctuations. From the simplistic point of view of the Bohr model of an atom, in which electrons move in orbits round the nucleus, such fluctuations could be considered to be related to the movement of the electrons. From the point of view of quantum mechanics, it turns out that such fluctuations, that result in a fluctuating dipole moment, are related in both frequency and magnitude to the spectral absorption bands of the molecule.

The fluctuating dipole propagates a fluctuating electric field which moves outwards with the velocity of light. This fluctuating field induces a fluctuating dipole in any molecule it encounters. This second fluctuating dipole then radiates a fluctuating field back to the first molecule where it interacts with the original fluctuation. The second molecule behaves in the same way with respect to the first, so that the effect is mutual. The net effect of these interactions is to produce a negative energy of interaction and an intermolecular force which is invariably attractive. Such forces are usually termed London dispersion forces after London who first described them; they are also sometimes called van der Waals forces since they play an important role in the theory of non-ideal gases. It should be noted however that the term 'van der Waals force' is sometimes taken to include multipole interaction forces and induction forces as well.

A detailed investigation of the origin of dispersion forces shows that the energy of interaction depends on the absorption spectra of both participating molecules, and is proportional to the product of their polarisabilities and the inverse sixth power of their separation.

An expression for the energy of interaction of two identical molecules due to dispersion interactions is given in column 5 of row 6(*a*) of table 2.1. α_1 and α_2 are the polarisabilities of the molecules and $h\nu$ their ionisation energies. The corresponding entries in rows 6(*b*) and 6(*c*) give the energy of interaction of two water molecules at a separation of 0.3 nm and the separation required to give an energy equal to RT. The ionisation energy of water is taken as 12.6×10^{-19} J per molecule. It may be noted that energy of interaction due to dispersion interactions is comparable with that due to dipole-quadrupole interactions. An interesting effect arises if the two molecules are far apart. Since the electric fields are propagated with the velocity of light, c, the time taken for the first molecules to experience the field reflected from the second is $2r/c$. If r, the separation distance, is large enough, the high frequency components of the reflected field may be out of phase with the original fluctuation after this time delay. The net result is that the high frequency contributions to the interaction energy are damped, and the dispersion forces are said to be retarded. It is found that in this case the energy is proportional to the inverse seventh power of the separation. It should be noted, however, that this only occurs at such large distances that the interaction energy is in any case small.

2.5   Exclusion forces

If the valencies of two molecules are both fully utilised so that no further covalent bonds are possible (we continue to assume that this is the case) and they are brought close together, they experience a strong repulsive force when their electron distributions start to overlap. This can be seen to result from the mutual repulsion of the two electron distributions but in a deeper sense is a quantum mechanical effect which is a consequence of the Pauli exclusion principle. As the two molecules are brought still closer together the intermolecular energy rises steeply. An approximate treatment suggests that it depends on the inverse twelfth power of the separation. This implies that there is an almost perfectly sharp energy barrier which prevents two atoms approaching each other closer than a critical distance. This critical distance depends on the two atoms concerned and is, to a good approximation, the sum of two 'radii', each characteristic of one of the atoms. These radii are termed the *van der Waals radii*. A further consequence is that each atom and molecule is associated with a volume which other atoms cannot enter. This is the *excluded volume* of the atom or molecule.

We finally see that the total intermolecular energy can be expressed as the sum of a number of terms of the form Ar^{-n}, one for each type of interaction that is operating. The constants, A, depend on the nature of the molecules and on the nature of the corresponding force; the exponents, n, depend only on the latter. In particular, for the interaction of uncharged, non-polar molecules, only terms in r^{-6} and r^{-12} are necessary. In this case the total interaction energy may be expressed in the form

$$V(r) = 4 V_e \left[\left(\frac{r}{r_0} \right)^{-12} - \left(\frac{r}{r_0} \right)^{-6} \right] \tag{2.2}$$

where r_0 is the separation at which the energy becomes zero and V_e is the minimum energy. A potential-energy relation taking this form is known as a Lennard–Jones potential. Following this relation, $V(r)$ is plotted against r in figure 2.1.

2.6   Interactions in solution

So far we have tacitly assumed that the interacting molecules are isolated and in a vacuum. We must now consider the consequences that ensue if they are immersed in a solvent. Several different effects need to be considered.

First let us be clear that we are here concerned with the total energy change when two solute molecules (or monomer units) already immersed in the solvent are brought to a separation distance, r, from a great distance. We shall assume that this energy change is negative.

The first effect to take note of depends upon the fact that, from their energy of interaction, estimated in the manner described above and taking into account other effects discussed below, must be subtracted the energy of interaction of the two portions of solvent that they displace when at the separation, *r*. This 'Archimedes' effect results in a diminution of the energy of interaction.

A second and minor effect arises from the fact that the velocity of light in a medium is less than its value *in vacuo*. This means that the dispersion forces will become retarded at shorter separations than they do *in vacuo*. This too is expected to diminish the interaction energy.

The expression for the force between two point charges given by Coulomb's law requires modification if the charges are surrounded by a dielectric (the solvent). This modification requires that the permittivity of free space, ϵ_0, which occurs in the various expressions for the interaction energies, should be multiplied by ϵ_r, the relative permittivity of the medium (otherwise known as the dielectric constant) to give the permittivity of the medium. In order that the reader should not lose sight of the factor ϵ_0, we will continue to write the permittivity of the medium as $\epsilon_r\epsilon_0$. The relative permittivity of water at 298 K is 78 so that all the energies given in table 2.1 would be reduced by this factor if the interacting molecules were surrounded by water. It is useful to note that the product, $\epsilon_r T$, is approximately constant (23 200 K) for liquid water.

From a molecular standpoint, the reduction of the electrostatic energy of interaction of two particles is a consequence of the interaction of the particles with the molecules of the dielectric. The electric field due to the interacting particles polarises the dielectric; it does this in the first place by inducing dipoles in the molecules of the dielectric, and in the second place by causing any permanent dipoles they may carry to orientate in the direction of the field, an effect opposed by random thermal motion. Energy is required to bring about this polarisation of the dielectric, and this energy may be said to be stored in the dielectric. When the two interacting particles approach one another some of this stored energy is liberated. It is this liberation of energy which accounts for the decrease in energy of interaction between the particles in a dielectric.

Two points may be noted. Firstly the effect of the dielectric will still be present even if the particles are so close together that there is no room for solvent molecules between them. Secondly the effective dielectric constant in the vicinity of a particle may not be uniform; this is particularly the case in the vicinity of ions which strongly bind water molecules so that they are effectively immobilised.

If the two particles approach each other very closely, any strongly bound solvent which would otherwise separate them must be removed first. The energy required to do this must be included in the total energy of interaction.

It should be clear from these remarks that an exact computation of the energy of interaction in a dielectric medium or solvent is difficult. In the first place there is uncertainty as to the correct value of the dielectric constant that should be used when the interacting particles are close together; the value of the bulk dielectric constant may be inappropriate. On the other hand an attempt to take into account the interactions with all the solvent molecules using the molecular approach is prohibitively difficult.

Nevertheless several general points emerge. Firstly, the interaction energy is reduced by the presence of the solvent. Secondly, this reduction will be more marked in the presence of polar solvent molecules than in the presence of non-polar molecules. Thirdly, the interactions of particles that interact strongly with the solvent will be inhibited.

Finally, we may note that solvents whose molecules are strongly polar (such as water) are termed *polar solvents;* those whose molecules are non-polar (such as cyclohexane) are termed *non-polar solvents.*

2.7 The hydrogen bond

A particularly important type of secondary interaction is the *hydrogen bond.* This plays an important role in stabilising the tertiary structure of globular proteins and the helical secondary structures encountered in polypeptides and polynucleotides.

The formation of a hydrogen bond requires two electronegative atoms such as N, O or F. One of these must be covalently bonded to a hydrogen atom; as such it constitutes the donor group. The other constitutes the acceptor group.

The bond to the hydrogen atom in the donor group is strongly polarised as we have suggested in section 2.2 and is thus associated with a dipole moment which is aligned along the bond and points towards the hydrogen atom. The acceptor group is also associated with a dipole moment which results from the separation of the unpaired electrons from the electronegative atom. It is aligned in a direction determined by the position of their orbitals and points towards the nucleus of the atom. These two dipoles can enter into a strong dipole-dipole interaction which is the primary source of the energy of formation of the hydrogen bond.

We have seen in section 2.2 that dipole-dipole interactions are at their strongest when the two dipoles are aligned head to tail. This implies that the two electronegative atoms and the hydrogen atom will tend to be colinear. However, deviations from linearity of up to $20°$ result in only a 12% diminution of the dipole-dipole interaction energy and deviations of this magnitude are frequently observed in hydrogen bonds in crystals as observed by X-ray diffraction techniques.

The unique character of hydrogen bonds and the reason for singling them out from other dipole-dipole interactions is associated with the very small van der Waals radius of the hydrogen atom. This implies that the two dipoles can approach each other very closely, with a corresponding enhancement of the interaction energy and the force between them. So strong does this force become at such short ranges that the distance between the nuclei of the hydrogen atom and acceptor group is commonly observed to be about 0.02 nm less than the sum of their van der Waals radii.

It should be noted that, although the interaction energy is predominantly that of the dipole-dipole interactions, significant contributions also arise from other multipole interactions and from induction and dispersion forces. It is also possible that a small contribution arises from quantum mechanical effects leading to the hydrogen bond taking on a partially covalent character.

The energy of formation of a hydrogen bond, V_e, is characteristically of the order of 20 kJ/mol.

2.8   The structure of water

Water is the solvent in which most biochemical reactions occur and the solvent in which the properties of proteins and nucleic acids are most usually investigated.

Water also has certain unique properties both as a liquid and as a solvent. These arise from the fact that the water molecule, H_2O, can act both as a donor and as an acceptor in the formation of hydrogen bonds. The electrons in the outer shell of the oxygen atom are in sp^3 hybridisation so that the two OH bonds and the two pairs of unshared electrons are directed towards the corners of a tetrahedron. Each unshared pair of electrons is displaced from the nucleus to give a dipole component directed towards the nucleus; similarly each OH bond carries a dipole directed away from the oxygen nucleus. Thus each water molecule can donate a proton to two hydrogen bonds and also accept two protons from two separate hydrogen bonds.

It is thus possible for an assembly of water molecules to link up and form a lattice in which each molecule is hydrogen bonded to four others in a tetrahedral arrangement similar to the structure of diamond. Such a structure is observed in crystals of ice I. Crystal structures with variations on this theme are found in the other crystalline forms of ice.

One notable feature of ice I is its low density. This is attributable to the open nature of the lattice. Indeed, the interstitial cavities between the water molecules (cavities marked by oxygen atoms at the corners of an empty tetrahedron) are sufficiently large to accommodate a water molecule.

Although the density of water is greater than that of ice, there is abundant evidence that water itself retains some structure and that the individual molecules are extensively hydrogen bonded and not randomly arranged in the liquid. The precise nature of this structure is still a matter of controversy, and none of the models that have been proposed are consistent with all of the experimental data.

The most widely quoted and perhaps the most highly elaborated model is that of Frank and Wen as further developed by Nemethy and Scheraga. In this 'flickering cluster' model, the individual water molecules tend, by a cooperative process, to form into structured clusters which resemble, but are not necessarily identical to, the hydrogen-bonded lattices found in ice crystals. It is thus supposed that liquid water consists of innumerable clusters of varying sizes which are in equilibrium with each other and also with free unstructured water molecules. The integrity of one of these clusters depends on the simultaneous presence of several hydrogen bonds between its constituent molecules. If, as a result of a local fluctuation of the energy, one or more of these hydrogen bonds is disrupted, the whole cluster tends to collapse. Conversely clusters are continually forming from the free water molecules as a result of similar fluctuations. In this way the clusters rapidly form and collapse or 'flicker'.

As we shall see below, the introduction of a solute molecule into water may affect the structure of the latter. Some solutes tend to disrupt the clusters and some tend to facilitate their formation.

2.9   Solvent–solute interactions

Let us consider a very dilute solution, so dilute that the molecules of the solute rarely encounter each other. Each solute molecule is nevertheless surrounded by solvent and interacts with it through the mediation of the various forces described above. A measure of this interaction is provided by the standard term in the expression for the chemical potential. We shall discuss this thermodynamic parameter in more detail in the next chapter and will only note here that it includes the free energy change when one mole of solute molecules is dissolved in a very large (infinite) volume of solvent. As such it contains a term corresponding to the energy change and a term corresponding to the entropy change.

Electrically charged ions in a strongly polar solvent such as water tend to be solvated or hydrated; that is to say they tend to bind solvent or water molecules on account of strong monopole–dipole interactions. The smaller the ion, the stronger the electric field at its periphery, and hence the more firmly are bound the solvent molecules and the greater their number. If solvent molecules are strongly bound in this way they are localised, and this implies a loss of entropy; but this loss is partly compensated if the bound water is obtained from the breakdown of water structure.

A similar situation prevails in the case of polar solute molecules in a polar solvent, though in this case the interaction, which depends mainly on dipole–dipole interactions, is not nearly so strong. Nevertheless, if the solute carries hydrogen-bond donor or acceptor groups and the solvent, like water, is of a similar nature, it is expected that the solvent will be extensively hydrogen bonded to the solute. If the solvent is non-polar, the only interactions with a polar solute will be weak induction and dispersion interactions. These will also be the interactions which occur between a non-polar solute and a non-polar solvent.

It might be expected that these weak residual forces would be the only ones operating between a non-polar solute in a polar solvent. If, however, the polar solvent is water, an interesting new effect arises, and this will be discussed in the next section. It is important to note however that before any solute molecule can be inserted into the body of the solvent, a hole must be formed to contain it. If the solvent molecules interact strongly between themselves, energy is required to do this. This energy is included in the standard free energy change.

We thus see that the total standard free energy change contains contributions to both its energy and entropy components which arise from various sources. Moreover, these different contributions may be of different sign, and each is difficult to estimate with precision. It is therefore difficult to assess even the sign of this quantity in any particular case.

One important point about the standard free energy is that it is crucial in determining the solubility of the solute. Let us consider a saturated solution of the solute which is in equilibrium with the pure solute and such that the mole fraction of the solute in the solution is x_2. At constant temperature and pressure, the laws of thermodynamics tell us that the chemical potential (see chapter 3) of the solute in the solution is equal to the molar free energy of the pure solute. If we continue to assume (for simplicity) that the solution is dilute, so that interactions between the solute molecules therein can be ignored, this implies that

$$\mu_2^* = \mu_2^\ominus + RT \ln x_2 \tag{2.3}$$

where μ_2^* is the molar free energy of the pure solute and $\mu_2^\ominus$ is the standard chemical potential of the solute in the solution. We thus see that the solubility, expressed as the mole fraction of the solute in the saturated solution, may be expressed as $e^{[-(\mu_2^\ominus - \mu_2^*)/RT]}$. We have indicated above the difficulties attendant on estimating $\mu_2^\ominus$. Similar difficulties are involved in assessing μ_2^*, which depends on intermolecular interactions in the pure solute. Nevertheless, certain qualitative generalisations emerge.

Firstly, electrolytes will tend to be appreciably dissociated only in highly polar solvents such as water. This is because the energy required to separate two strongly interacting ions must be compensated by strong interactions with the solvent.

For similar reasons polar solutes tend to be appreciably soluble only in polar solvents. Conversely non-polar solutes tend to be soluble in non-polar solvents.

2.10 The hydrophobic bond

It is well known that oil is immiscible with water; this is believed to be a consequence of an effect alluded to in the previous section.

Let us consider the process of transferring a non-polar molecule such as hexane (to take a simple example) from pure liquid hexane into aqueous solution. Hexane is non-polar so that only weak intermolecular interactions occur between the molecules in the pure liquid hydrocarbon, and little energy is required to remove our molecule. Similarly the hexane will interact but weakly with the water molecules that surround it in the solution.

Nevertheless, it is found from experiment that the saturation solubility of hexane in water is only 1.8×10^{-4} mol/l at 303 K which corresponds to a value of $\mu_2^* - \mu_2^\ominus$, that is the free energy change accompanying the transfer of one mole of hexane from pure hexane to solution, equal to +33 kJ/mol. This is too large to be accounted for in terms of the weak induction and dispersion interactions mentioned above. It is also found from a study of the effect of temperature on the solubility that both the enthalpy (or energy) change and the entropy change accompanying the transfer are both negative, with the latter dominating the former at 300 K. This implies that the solubility decreases as the temperature is raised.

This decrease in the entropy has been attributed to changes in the structure of the water. In terms of the flickering cluster model, this has been rationalised along the following lines. We have argued that liquid water consists of structured clusters of water molecules in equilibrium with non-bonded molecules in the spaces between the clusters. A non-polar solute molecule placed in one of these spaces amid the non-bonded molecules interacts with them less strongly than they interact among themselves, but disrupts hydrogen bonds between them. This raises the energy of the non-bonded water molecules and so diminishes the probability of their occurrence, with the consequence that they tend to associate with the clusters. Thus the clusters tend to grow at the expense of the free water molecules, and this process is accompanied by a decrease in the energy and of the entropy as the free water molecules form hydrogen bonds in the clusters and become localised.

Whether or not explanations along these lines are ultimately correct, the fact remains that non-polar solutes are but sparingly soluble in water. This is equivalent to saying that non-polar solutes tend to aggregate among themselves rather than become dispersed in water. This has given rise to the concept of a hydrophobic bond between the non-polar molecules in the presence of water. This is a useful shorthand way of referring to the effect, provided it is clearly understood that there is no actual force (or bond) acting between the aggregated non-polar molecules and that the effect originates from changes in the thermodynamic properties of the water.

Hydrophobic bonds are believed to be of great importance in stabilising the tertiary structure of globular proteins and in stabilising micelles and other structures discussed in section 1.3 and illustrated in figure 1.3. In these, it will be observed, the polar part of the detergent or lipid molecule is exposed to the aqueous solvent and the non-polar parts are aggregated together. Such molecules with both polar and non-polar parts are sometimes called *amphiphilic*.

2.11   Intramolecular interactions in polymers

The interaction between two parts (monomer units) of the same polymer chain is governed by the same considerations that govern intermolecular interactions and that we have discussed above. Thus two units each carrying polar groups will tend to interact; two charged groups on a polyelectrolyte will attract or repel each other according to whether their charges are of opposite or the same sign; a hydrogen-bond donor group is able to bind to a hydrogen-bond acceptor group on the same polymer molecule; two monomer units will be prevented from occupying the same volume by exclusion forces; hydrophobic groups on the same polymer chain dissolved in water will tend to aggregate. Similarly, the monomer units will interact with the solvent in the same manner as other molecules as discussed in section 2.9.

In principle, any one of the many monomer units may interact with any other and, in general, two interacting units will be widely separated along the polymer chain. Such interactions are termed *long range interactions* and these will be discussed in greater detail in the next two chapters.

Of equal importance are short range interactions between neighbouring parts of the chain since these intimately affect the flexibility of a polymer chain.

2.12   Bond rotation potentials

Let us first consider a simple molecule such as ethane (CH_3-CH_3). One of the methyl groups may rotate relative to the other about the carbon-carbon bond. As this happens the energy of the molecule changes and, in a complete revolution, passes through three minima and three maxima as indicated in figure 2.2. The minima correspond to staggered configurations and the maxima to eclipsed configurations as depicted in figure 2.3, which shows in diagrammatic form (Newman projections) views looking along the carbon-carbon bond. It is possible, by spectroscopic means, to measure the energy difference between these two extreme forms and this turns out to be 12.2 kJ/mol. The origin of this energy difference is not entirely beyond controversy, but secondary interactions and exclusion forces between the hydrogen atoms on the two methyl groups play a role.

The figure of 12.2 kJ/mol is 4.9 times the value of RT at 300 K. The Boltz-

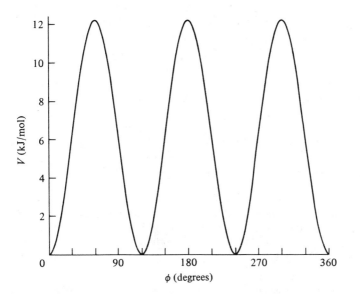

Figure 2.2. Rotational potential energy diagram for ethane showing the manner in which the energy varies as the rotation about the carbon–carbon bond changes. The angle ϕ is taken as zero for the staggered form, and the energy difference between the eclipsed and staggered forms is taken as 12.2 kJ/mol. The exact form of the relation between ϕ and V is unknown, and the curve has been plotted on the assumption that $V = \frac{1}{2} V_0 (1 - \cos 3\phi)$, with $V_0 = 12.2$ kJ/mol.

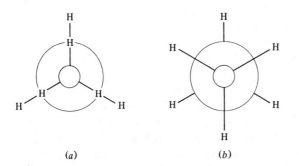

(a)   (b)

Figure 2.3. Newman projections of ethane: (a) eclipsed configuration; (b) staggered configuration. The Newman projection represents a view along the carbon–carbon bond; the two circles represent the two carbon atoms with their pendant hydrogen atoms.

mann distribution then implies that configurations close to the staggered form are statistically favoured, and a rough calculation suggests that there is a 98% chance of finding an ethane molecule in the staggered form. Notwithstanding this, it is also clear that a molecule will frequently acquire sufficient

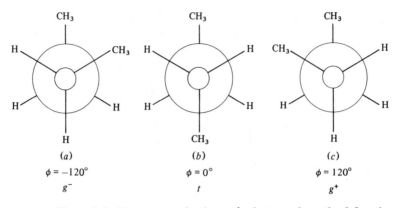

Figure 2.4. Newman projections of n-butane along the 2,3-carbon–carbon bond: (a) *gauche* configuration g^-; (b) *trans* configuration t; (c) *gauche* configuration g^+. The angle ϕ is defined to be zero for the *trans* configuration.

energy to enable it to rotate from one staggered form, through an eclipsed form, to another staggered configuration. This is expected to occur at a rate of about 10^{10} times a second. It is important to note that, since ethane is symmetrical, the potential energy diagram in figure 2.2 has three-fold symmetry.

This three-fold symmetry is lost in n-butane ($CH_3-CH_2-CH_2-CH_3$). If we consider rotation about the 2,3-carbon–carbon bond we still expect there to be staggered and eclipsed configurations, but now there are two sorts of staggered forms, as illustrated in figure 2.4. There is one, termed the *trans* configuration, in which the 2,3-carbon–carbon bond and the two methyl groups are in the same plane and there are two in which they are not; these are termed the *gauche* configurations. For brevity the *trans* is designated as a t state and the two *gauche* as g^+ and g^- states. The potential-energy diagram now resembles that shown in figure 2.5 and there it may be seen that the energy minimum corresponding to the t state is lower than that of either of the two g states. The energy of the t state is about 2.1 kJ/mol lower than that of the two g states, but the barrier between the minima is about 15 kJ/mol. This means that rotation from one minimum to another is still a frequent event, but the t state is statistically favoured; the probability of observing a t state (as opposed to either g state) is about 66%.

If we now consider n-pentane ($CH_3(CH_2)_2CH_3$) and examine rotations about both the 2,3-carbon–carbon bond and the 3,4-carbon–carbon bond we see that there are nine states defined by the configuration of these two bonds as tt, tg^+, tg^-, g^+t, g^-t, g^+g^-, g^-g^+, g^+g^+ and g^-g^-, with several different energy levels. If these forms are examined in detail it is found that the two

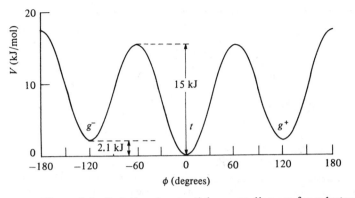

Figure 2.5. Rotational potential energy diagram for n-butane. ϕ is taken as zero for the *trans* configuration. The difference between the energy of the *trans* and *gauche* minima is taken as 2.1 kJ/mol (ΔV_{tg}) and the height of the energy barrier between the *trans* and *gauche* states has been taken as 15 kJ/mol (ΔV_h). The exact form of the relation between V and ϕ is unknown. The diagram has been plotted on the basis of the assumption that it is given by $V = \frac{1}{2}(\Delta V_h - \frac{1}{3}\Delta V_{tg})(1 - \cos 3\phi) + \frac{2}{3}\Delta V_{tg}(1 - \cos \phi)$. The value of the barrier between the two *gauche* states given by this equation may be seriously in error.

terminal methyl groups of the g^+g^- and the g^-g^+ states are in close proximity, so close that exclusion forces considerably raise the energy of this configuration and render it statistically improbable.

Similar considerations apply to higher homologues of these hydrocarbons including polymethylene. With each carbon–carbon bond there is associated three energy minima, but the actual values of the energy at the minima depend on the state of the neighbouring bonds. Indeed, similar considerations apply to all polymers in which there is the possibility of rotation about the bonds linking the atoms of the backbone. The number of minima, the angular displacements at which they occur and their energies depend on the precise structure of the polymer.

In order to assess the relative probabilities of the various rotational states it is necessary to know the energies associated with them. In general the energy associated with rotation about a particular bond depends on two angles: the angle ϕ_2 defining rotation about the bond in question and also ϕ_1 which defines the rotation about the preceding bond. It is convenient to depict this dependence in the form of contours of equal energy in a plane whose coordinates define value of ϕ_1 and ϕ_2. Such a conformational energy map is shown in figure 2.6. In order to construct such maps a semi-empirical approach must be used.

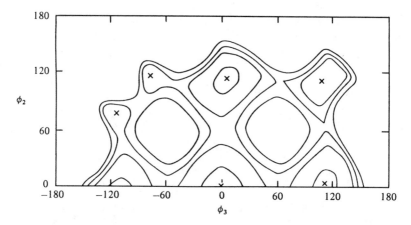

Figure 2.6. Conformational map for n-pentane. This gives contours of equal energy for rotations about the 2,3-carbon–carbon bond and about the 3,4-carbon–carbon bond. Contours are shown at intervals of 4.2 kJ/mol and minima are indicated by X. From P. J. Flory, *Statistical Mechanics of Chain Molecules*. (Interscience, New York, 1969).

As ϕ_1 and ϕ_2 vary, the different atoms and side groups attached to the backbone atoms involved change their relative positions. This implies that the energy of interaction between them changes. In principle one may write down an expression which gives the sum of the interaction energies between all pairs of pendant groups in terms of ϕ_1 and ϕ_2, and this could be used to plot the energy contours. The practical application of this approach requires information concerning the dipole strengths, polarisabilities and so forth of all the groups concerned. Such information can sometimes be obtained from the study of simple molecules. It should be noted that, as well as the interaction energies between the groups, it is usually necessary to assume that there is an intrinsic variation of energy with rotation about a bond. This probably arises from quantum mechanical effects which may make the bond acquire a partial double bond character. Again relevant parameters must be gleaned from experiments on simple compounds.

Conformational energy maps constructed in this manner are open to inaccuracy and error, and the relevant parameters are available for some polymers with a greater precision than for others. In chapter 4 we shall see how such maps are utilised in elucidation of the statistical properties of polymers.

Problems

2.1   What are the dominant intermolecular forces acting between the following pairs of molecules or groups?
    (a) The side groups of glycine and serine

(b) The side groups of glycine and lysine
(c) The side groups of lysine and tyrosine
(d) Two methylene groups
(e) The ionic forms of uridylic acid and adenylic acid
(f) Two molecules of ribose
(g) Water and uracil
(h) Uracil and adenine

2.2 Which of the amino acids would you expect to find at the surface of a globular protein and which in the interior of the molecule when in its native tertiary structure.

2.3 In the secondary structure of DNA two polynucleotide chains are in close proximity with hydrogen bonds linking a base in one chain with a complementary base in the other to form a base pair. The phosphate groups carried by the two nucleotide residues participating in a base pair are about 2 nm apart. Calculate the magnitude and sign of the monopole interaction energy between them assuming each carries a single electronic charge. Is this an underestimate or an overestimate of the total electrostatic interaction energy per base pair for the two long polynucleotide chains. Assume a dielectric constant of 5.

2.4 Compare the energy calculated in problem 2.3 with the hydrogen-bond energy per base pair assuming that two hydrogen bonds are involved in each pair. What conclusion do you draw?

2.5 How would the monopole interaction energy calculated in problem 2.3 be changed if the two counterions associated with each phosphate group were localised at a distance of 2 nm from the phosphate group, on the straight line joining them and in the order $X-P-P-X$ (X is a counterion and P a phosphate group).

2.6* A dipole may be considered to be defined by two charges $+Q$ and $-Q$ separated by a distance x. The dipole strength is Qx. Derive an expression for the interaction energy between a monopole, charge Q_m, and the dipole in terms of the dipole strength, p, the charge, Q_m, the distance between the charge and the centre of the dipole, r, and the angle between the line joining the charge to the dipole and the direction of the dipole. Assume $x \ll r$.

2.7 Table 2.1 gives values of various intermolecular interaction energies calculated for separations of 0.3 nm. Recalculate similar values for separations of 0.9 nm (note: it is not necessary to work out each expression afresh).

2.8 Suggest reasons why acetic acid is soluble in water but dodecanoic acid is not, and why sodium chloride is soluble in water but not in benzene.

3 The equilibrium thermodynamics of solutions of large molecules

Classical thermodynamics is concerned with the macroscopic properties of a system and the relations that must hold between them at equilibrium. In this chapter we shall be concerned with the thermodynamic properties of solutions of large molecules.

We shall, however, also be concerned with the relation between these macroscopic properties and the molecular properties of the macromolecular solute. To this end we shall invoke some of the principles of statistical thermodynamics.

In this way we shall attempt to relate measurable properties of solutions of large molecules to their molecular structure. Such relations not only provide a rationalisation of certain unique properties of such solutions, but also enable us to glean information concerning molecular architecture from experimental measurements conducted at equilibrium.

Before we embark upon this programme, we shall briefly review some relevant aspects of thermodynamic theory and terminology.

3.1 Extensive, intensive and partial properties

Any property of a system whose value is equal to the sum of the values of the different parts of the system is an *extensive* property. Examples are mass, m, volume, V, energy, U, entropy, S, enthalpy, H, or, G, Gibbs free energy, to name but a few.

Properties which do not have this characteristic are *intensive* properties such as pressure, P, temperature, T, or chemical potential. It makes no sense to say, for instance, that the temperature of a system is the sum of the temperatures of its parts; rather, the temperature is uniform throughout the system at equilibrium.

For each extensive property, there is a corresponding partial molar and a partial specific property. Consider the change in an extensive property, Y, when a small quantity of one of the components, i, is added to the system, the temperature and the pressure and the quantities of all the other components being kept constant. If the quantity of the component i is measured in moles, the ratio of the change in Y to the quantity of i added, dn_i, defines the partial molar value of that component:

$$\overline{Y}_i = \left(\frac{\partial Y}{\partial n_i}\right)_{T,P,n_j} \tag{3.1}$$

If the quantity of the component is measured in units of mass, there results, by a similar definition, the partial specific value of Y for the component, $\overline{y}_i = (\partial Y/\partial m_i)_{T,P,m_j}$. Since the number of moles of i, namely n_i, is related to the mass present, m_i, by $m_i = M_i n_i$, where M_i is the molar mass of i (i.e. the mass per mole of i with dimensions of mass/quantity to be distinguished from the 'molecular weight', which is dimensionless), we have a general relation between partial specific and partial molar quantities:

$$\overline{Y}_i = M_i \overline{y}_i \tag{3.2}$$

If the system consists of a pure component we define corresponding properties termed the molar or specific quantity, written Y_i^* or y_i^* (see table 3.1), and defined as the total value of the property Y divided by the number of moles or the mass of the pure component present.

In general, the partial molar properties of the components in a system depend on the concentrations of all the components present as well as on the temperature and pressure, but they are not all independent. Let us suppose that small amounts, dn_1, dn_2 etc. of each of the components are added to the solution (P and T being kept constant). The total increase in Y is then given by

$$dY = \overline{Y}_1 dn_1 + \overline{Y}_2 dn_2 + \cdots \tag{3.3}$$

If the amounts of all the components added are all in the same proportion to the amounts already present, so that $dn_i = n_i d\xi$, we see that the increase in Y is $\overline{Y}_1 n_1 d\xi + \overline{Y}_2 n_2 d\xi + \cdots$. This expression may be integrated to give

$$Y = \overline{Y}_1 n_1 + \overline{Y}_2 n_2 + \cdots \tag{3.4}$$

We may now differentiate equation (3.4) to obtain the total differential of Y in the form $n_1 d\overline{Y}_1 + \overline{Y}_1 dn_1 + n_2 d\overline{Y}_2 + \overline{Y}_2 dn_2 + \cdots$ and subtraction of equation (3.3) from this yields

$$n_1 d\overline{Y}_1 + n_2 d\overline{Y}_2 + \cdots = 0 \tag{3.5}$$

This useful expression shows that small changes brought about in the values of the partial molar property by some agency are not independent. In particular, for a two-component system, $d\overline{Y}_2 = -(n_1/n_2) d\overline{Y}_1$.

Two partial molar properties are of particular interest to us, the partial molar volume and the partial molar Gibbs free energy, but before we discuss these we will briefly discuss the specification of the concentration of solutions.

Table 3.1. *Table of symbols for thermodynamic quantities*

I For any extensive property, Y:

$\overline{Y}_i$ Partial molar property of component i $(\partial Y/\partial n_i)$

$\overline{y}_i$ Partial specific property of component i $(\partial Y/\partial m_i)$

Y_i^* Molar property of pure component i (Y/n)

y_i^* Specific property of component i (Y/m)

II Some thermodynamic functions of state:

T Absolute temperature

P Pressure

V Volume

U Internal energy

H Enthalpy $(U + PV)$

S Entropy

A Helmholtz function (free energy) $(U - TS)$

G Gibbs function (free energy) $(U + PV - TS)$

III Quantities describing composition of solution or mixture:

n_i Amount of component i present (number of moles)

x_i Mole fraction of component i $(n_i/\Sigma_i n_i)$

m_i Mass of component i present

w_i Mass (weight) fraction of component i $(m_i/\Sigma_i m_i)$

c_i Concentration (moles/unit volume) of component i (n_i/V)

ρ_i Mass concentration of component i (m_i/V)

N_i Number of molecules of type i present $(n_i N_A)$

IV Other quantities pertaining to the components of a solution or mixture:

M_i Molar mass of component i (m_i/n_i)

μ_i Chemical potential of component i $(\partial G/\partial n_i)$

$\mu_i^{\ominus}$ Standard chemical potential of component i

λ_i Absolute activity of component i $(e^{\mu_i/RT})$

$\lambda_i^{\ominus}$ Absolute activity of component i in standard state $(e^{\mu_i^{\ominus}/RT})$

a_i Relative activity of component i $(\lambda_i/\lambda_i^{\ominus})$

f_i Activity coefficient of component i (mole fraction scale) (a_i/x_i)

y_i Activity coefficient of component i (concentration scale)
$[a_i(c)/c_i$; $a_i(c)$ is the activity measured on the concentration (moles/unit volume) scale]

V Miscellaneous constants:

N_A Avogadro constant $(6.022169 \times 10^{23}\ \text{mol}^{-1})$

R Gas constant $(8.31434\ \text{J K}^{-1}\ \text{mol}^{-1})$

k Boltzmann constant, R/N_A $(1.380622 \times 10^{-23}\ \text{J K}^{-1})$

3.2 The concentration of mixtures and solutions

The simplest and most fundamental way of specifying the relative proportions of the different components in a mixture or solution is by their *mole fractions*, x_i, defined by

$$x_i = n_i / \sum_i n_i \tag{3.6}$$

where n_i is the amount of component i present (usually specified in moles) and the summation extends over all components. It is clear from equation (3.6) that $\sum_i x_i = 1$. An alternative method is to specify the mass fraction, w_i, defined by

$$w_i = m_i \bigg/ \sum_i m_i = M_i n_i \bigg/ \sum_i M_i n_i \tag{3.7}$$

where m_i is the mass of component i present. Again, $\sum_i w_i = 1$.

It is often convenient and useful to specify concentrations, c_i, or weight concentrations, ρ_i, as the amount (usually in moles) or the mass of component i present in unit volume of the solution. It is readily shown that the various concentrations are related by

$$c_i = \rho_i / M_i = x_i \rho \bigg/ \sum_i x_i M_i \tag{3.8}$$

and that $\sum_i \rho_i = \rho$. Here ρ is the density of the solution. It must be recognised however that, in so far as ρ varies with temperature, so do ρ_i and c_i.

We shall often be concerned with two-component solutions and mixtures. We shall reserve the term *solution* for systems in which one of the components is present in excess. The component that is in excess is the *solvent* and the other(s) the *solute(s)*. We shall reserve subscript 1 for the solvent. If none of the components is present in obvious excess we shall use the term *mixture*. In this chapter we shall be concerned with mixtures of a macromolecular component with a low-molecular-weight component. We shall reserve subscript 1 for the latter. The difference between solutions and mixtures is a matter of convenience, and no unassailable distinction can be drawn between them.

We shall frequently be concerned with *dilute solutions* for which $m_1 \gg m_2$ and hence $n_1 \gg n_2$. For dilute solutions of one solute the following set of approximate relations may be assumed:

$$x_1 \simeq 1; \quad c_1 \simeq \rho_0 / M_1 = 1/V_1^*; \quad \rho_1 \simeq \rho_0$$
$$x_2 \simeq n_2 / n_1; \quad c_2 \simeq n_2 \rho_0 / n_1 M_1 = n_2 / n_1 V_1^*; \quad \rho_2 \simeq m_2 M_2 / n_1 V_1^*$$

$$\tag{3.9}$$

where ρ_0 is the density of the pure solvent and V_1^* the molar volume of pure solvent. We shall have frequent occasion to refer to these approximations. The symbols and their meanings for ways of specifying concentrations are gathered for reference in table 3.1.

3.3   Partial molar and partial specific volumes

The partial molar volume of a component is defined by equation (3.1) by replacing the general extensive property, Y, by the volume. From equation (3.2) we have $\bar{V}_i = M_i \bar{v}_i$.

For very dilute solutions, even those of macromolecules, the partial molar volume of the solute is generally found to be independent of concentration to a good approximation. This will be so if the solution is so dilute that the solute molecules are, on average, so far apart that they rarely encounter or interact with one another. Under these circumstances, the partial molar volume is equal to the increase in the volume of a large quantity of solution when one mole of pure solute is added. It would be tempting to equate this volume change to the volume from which solvent molecules are excluded by the presence of these extra solute molecules, a volume determined by the operation of the exclusion forces discussed in chapter 2. This temptation should be resisted because we must also take into account any change in the volume of any solvent that becomes bound to or otherwise interacts with the solute.

If we denote: by v_e the volume not available to solvent molecules because of the operation of exclusion forces due to the presence of one mole of solute; by m_s the mass of solvent bound to each mole of solute (or more exactly, the mass of any solvent bound such that it suffers a change in density); by ρ_0 the density of unbound (free) solvent; and by ρ_s the average density of bound solvent, we may write

$$\bar{V}_2 = v_e + m_s \left(\frac{1}{\rho_s} - \frac{1}{\rho_0} \right) + v_k \tag{3.10}$$

Here we have included an extra term, v_k, to represent a further small effect arising from the fact that even a minute particle of negligible size excludes solvent from a finite volume. It does this by virtue of its thermal motion, which causes it to continually nudge solvent molecules out of a small volume in its vicinity (it might be called the 'angry man in a crowd' effect).

The magnitude of m_s depends on the nature of the solvent and of the solute and on the interactions between them. It is likely to be greater for polar solutes in polar solvents than for non-polar solutes in non-polar solvents. The significance of the effect depends on the difference between ρ_s and ρ_0. This difference is likely to be negligible for unstructured non-polar solvents, but in water it may be large. The density of water is abnormally low on account of its structure as discussed in section 2.8. When water is bound to a polar or ionic solute, this structure tends to collapse. It may then turn out that for small solute molecules the decrease in volume contingent on the collapse of the structure of bound solvent is greater than the volume represented by v_e. This is exemplified by lithium hydroxide which in fact has a negative partial molar volume. Globular protein molecules are believed to carry a layer of bound water at their surface, but the magnitude of this and its density are hard to estimate with precision.

At higher concentrations, the partial molar volume of a solute may be found to vary with the concentration. This will occur if the solute molecules interact with each other with a change in volume. There is likely to be a change in volume if such interactions involve a change in solvation.

Whether or not the partial molar volume or partial specific volume depends on the concentration, the density of the solution will (in general) change as more and more solute is added to it. It is possible to express the partial specific volume at any concentration in terms of the rate of change of density, ρ, with mass concentration, $d\rho/d\rho_2$, and the concentration ρ_2 by

$$\bar{v}_2 = (1 - d\rho/d\rho_2)/(\rho - \rho_2 \, d\rho/d\rho_2) \tag{3.11}$$

Thus, if the density of a solution of a solute is measured at a series of different concentrations, and these are plotted against the concentration, the partial specific volume is easily found as illustrated in figure 3.1. It may be noted that the denominator in equation (3.11) is equal to ρ' defined in that figure. If such a plot of density versus concentration is linear, with a slope $\Delta\rho/\rho_2$, the partial specific volume is independent of concentration and given by $(1/\rho_0)[1 - \Delta\rho/\rho_2]$. Under these circumstances we also see that the quantity $(1 - \bar{v}_2 \rho_0)$ is equal to $d\rho/d\rho_2$.

The density of solutions may be measured with great accuracy to give accurate values of partial specific volumes in this way. A modern method of measuring densities is to fill a small tube with the relevant solution and to set this vibrating at its resonant frequency. This frequency depends on the mass of the tube filled with solution and hence on the density of the latter. The period of vibration may be determined with great accuracy by electronic counting devices. It is thus possible to measure densities to one part in 10^7 on a routine basis.

3.4 Chemical potentials and activities

The chemical potential of a component in solution may be defined as the partial molar Gibbs free energy

$$\mu_i = (\partial G/\partial n_i)_{P, T, n_j} \tag{3.12}$$

The importance of this quantity may be seen as follows. Consider a system composed of a number of phases each open to a component i. One of the consequences of the second law of thermodynamics is that, if such a system

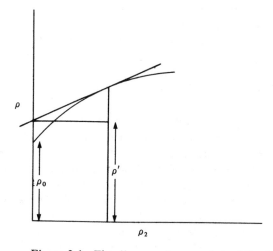

Figure 3.1. The diagram shows a plot of the density of a solution, ρ, versus the concentration of the solute. ρ_0 is the density of the pure solvent. The concentration of the solute, ρ_2, is defined as the mass of solute per unit volume of the solution. The partial specific volume of the solute, $\bar{v}_2$, at any concentration is given by

$$\bar{v}_2 = (1 - s)/\rho'$$

where s is the slope of the tangent at the corresponding point on the curve and ρ' is the intercept of the tangent with the ρ axis. If the partial specific volume is independent of the concentration, the tangent is coincident with the curve and $\bar{v}_2 = (1 - s)/\rho_0$.

is maintained at constant temperature and pressure, and is at equilibrium, the Gibbs free energy must be at a minimum. Let us suppose that dn_i moles of component i are transferred from phase I to phase II in which the chemical potentials are μ_i^I and μ_i^{II}. The loss of Gibbs free energy of phase I is $\mu_i^I dn_i$ and the gain to phase II is $\mu_i^{II} dn_i$. Thus the total net gain of both phases taken together is $\mu_i^{II} dn_i - \mu_i^I dn_i$. This must be zero if the free energy is to be at a minimum. It then follows that $\mu_i^I = \mu_i^{II}$ so that the chemical potential of any component which is in equilibrium with any two phases must be the same. This result can be generalised to any number of phases; at equilibrium the chemical potential of any component that can pass between them must be the same in each.

Unlike partial molar volumes, partial molar Gibbs free energies cannot be measured absolutely. All that it is possible to do is to measure the difference in chemical potential of a component between two solutions or states. One of these states may be selected quite arbitrarily, with an eye only to convenience, and termed the *standard state*. The chemical potential in the standard state of component i is written $\mu_i^\ominus$.

It is convenient to define another quantity in terms of the chemical potential. This is the *absolute activity*, λ, which is defined by

$$\mu_i = RT \ln \lambda_i \tag{3.13}$$

A similar definition defines the absolute activity in the standard state, $\lambda_i^{\ominus}$. It is then a trivial exercise to deduce that

$$\mu_i = \mu_i^{\ominus} + RT \ln (\lambda_i/\lambda_i^{\ominus}) = \mu_i^{\ominus} + RT \ln a_i \tag{3.14}$$

Here we have introduced yet another quantity, the *relative activity*, a_i (usually referred to as the *activity* for short), of the component i, as $\lambda_i/\lambda_i^{\ominus}$.

We have seen that the chemical potentials of each component in a system at equilibrium at constant T and P must be the same throughout. The same is true of the activities.

If we apply equation (3.5) to partial molar free energies, we see that these are not all independent and that

$$n_1 \, d\mu_1 + n_2 \, d\mu_2 + \cdots = 0 \tag{3.15}$$

This expression is known as the Gibbs–Duhem relation. When applied to a two-component system it enables the chemical potential of one component to be calculated as a function of composition, provided that of the other is similarly known as a function of composition.

The standard states of mixtures and of solutions are usually chosen in different ways. For mixtures it is convenient to define the standard states of each component as the pure component. The chemical potential of a pure substance is equal to its molar free energy, G_i^*. It thus follows from this specification of standard state that, for a mixture, $\mu_i^{\ominus} = G_i^*$.

For solutions the standard state of the solvent is defined in the same way so that $\mu_1^{\ominus} = G_1^*$, the molar free energy of the pure solvent. It turns out to be more convenient, however, to define the standard states of the solutes as their state at infinite dilution. At infinite dilution each solute molecule is a large (infinite) distance from every other one so that they do not interact with each other. Each molecule of solute is, so to speak, unaware of the presence of the others. It is important to note however that each solute molecule in the standard state is surrounded by solvent and is fully equipped with any solvation that the solvent–solute interactions may require and that such interactions are reflected in the value of $\mu_i^{\ominus}$. A difficulty arises in this definition of the standard state of the solutes in that their chemical potentials in this state of infinite dilution are, in fact, infinite. We shall deal with this difficulty in the next section.

It is important to note that the chemical potential of a component can also be expressed in terms of its partial molar enthalpy and entropy. By definition, the Gibbs free energy of a system is $G = H - TS$. It follows that

$$\mu_i = \overline{H}_i - T\overline{S}_i \tag{3.16}$$

A similar relation holds between the molar free energy, enthalpy and entropy, G_i^*, H_i^*, S_i^* of the pure components.

3.5   Ideal solutions

It will be recollected that the equation of state of an ideal gas takes a particularly simple form and that real gases behave as ideal ones at sufficiently low pressure. It is similarly convenient to define an ideal solution which obeys simple laws and such that real solutions behave like them at sufficiently high dilution (low concentration).

We thus define an ideal solution such that the chemical potentials of all components are given by

$$\mu_i = \mu_i^{\ominus} + RT \ln x_i \tag{3.17}$$

where x_i is the mole fraction of the component. Comparing this definition with equation (3.14) we see that the activity is equal to the mole fraction for an ideal solution. By putting x_i equal to unity we see that the standard chemical potential is equal to the molar free energy of the pure component for all components. Next we note from equation (3.4) that the total Gibbs free energy of a solution is given by $\sum_i n_i \mu_i$. We then see that the change in free energy that occurs when the components of any solution, originally in their pure standard states, are mixed to form the solution, as given by the free energy of mixing, ΔG_m, is

$$\Delta G_m = \sum_i n_i \mu_i - \sum_i n_i \mu_i^{\ominus} = RT \sum_i n_i \ln a_i \tag{3.18}$$

Volumes, ΔV_m, enthalpies, ΔH_m, and entropies, ΔS_m, of mixing are similarly defined. Since we have the thermodynamic relations, $V = (\partial G/\partial P)_T$ and $-H^2/T = (\partial (G/T)/\partial T)_P$, it follows that for an ideal solution

$$\Delta G_m^{id} = RT \sum_i n_i \ln x_i$$

$$\Delta V_m^{id} = \Delta U_m^{id} = 0$$

$$\Delta S_m^{id} = -R \sum_i n_i \ln x_i \tag{3.19}$$

In these expressions, we have written a superscript, id, to emphasise that the relations in which it occurs are valid only for ideal solutions. Equations (3.19) imply that the components of an ideal solution mix with no change of volume or of energy. It also follows that the energy of interaction of a solute molecule with the surrounding solvent in an ideal solution is the same as the energy of interaction of a solute molecule with its fellows in the pure state. This is of course not true for non-ideal solutions. The free energy of mixing is entirely accounted for by the entropy change.

We shall see later that a quantity of greater importance is the *relative chemical potential* of the solvent defined as $\mu_1 - \mu_1^\ominus$. For pure solvent, this is obviously zero. It only differs from zero when there is a finite concentration of some solute. The effect of solutes is to lower the chemical potential of the solvent, and it is convenient to express this relative chemical potential of the solvent as a power series in the concentration of the solute. For an ideal solution this may be done by noting from equation (3.17) that the relative chemical potential of the solvent is equal to $RT \ln x_1$. If we note that for a two-component solution $x_1 = 1 - x_2$, expand the logarithm as a power series in x_2 and then note that for dilute solutions $x_2 = \rho_2 V_1^*/M_2$ we obtain

$$\mu_1 - \mu_1^\ominus = -RTV_1^*\rho_2 \left[\frac{1}{M_2} + \left(\frac{V_1^*}{2M_2^2} \right) \rho_2 + \cdots \right] \qquad (3.20)$$

This equation gives the *virial expansion* of the relative chemical potential of the solvent in an ideal solution. For low concentrations we see that the relative chemical potential of the solvent approximates to $-RTV_1^*\rho_2/M_2$. The coefficient of the second term, $V_1^*/2M_2^2$, is the *second virial coefficient*.

Further insight into the significance of an ideal solution can be obtained from a consideration of the entropy of mixing. To do this we invoke the Boltzmann expression for the entropy of a system: $S = k \ln \Omega$, where Ω is the total number of distinguishable configurations (and assuming they all have the same energy). To apply this relation we suppose that the solution can be represented by a three-dimensional lattice consisting of a large number of cells. Into each cell may be placed either a solvent or a solute molecule. If there are N_1 solvent and N_2 solute molecules we suppose that there are $N_1 + N_2$ cells in all. We now estimate the number of ways, Ω, that we may place the N_1 solvent and the N_2 solute molecules in these cells. First we start placing the solute molecules in the empty lattice. The first may be placed in any one of the $N_1 + N_2$ empty cells; the second in any one of the remaining $N_1 + N_2 - 1$ cells, and so on. Thus the total number of ways of fitting in the solute molecules is $(N_1 + N_2)(N_1 + N_2 - 1) \cdots (N_1 - 1)$. This product of factors may be written in terms of factorials as $(N_1 + N_2)!/N_1!$. This result presupposes that the N_2 solute molecules are distinguishable. Since we suppose that they are all identical it does not matter in which order we place them in the lattice. We must therefore divide the result by the number of permutations of N_2 objects, namely $N_2!$. Next we must place the solvent molecules in the lattice, but since these too are all identical, they can only be fitted into the remaining N_1 cells in one distinguishable way. We thus see that the total number of configurations of the solution is $(N_1 + N_2)!/N_1!N_2! = \Omega$. Hence the entropy of the solution originating from the number of ways of mixing up the solvent and solute molecules is

$$S = k \ln [(N_1 + N_2)!/N_1!N_2!] \qquad (3.21)$$

For large numbers, such as are N_1 and N_2, the logarithm of $N!$ may be expressed by Stirling's approximation as $N \ln N - N$. It then follows from equation (3.21) and the definition of mole fraction that

$$S = -R(n_1 \ln x_1 + n_2 \ln x_2) \tag{3.22}$$

To obtain the entropy of mixing we must subtract from this the entropy of the pure solute and that of the pure solvent. In principle we apply the lattice model to each of these. In the first case we consider the number of configurations of N_1 solvent molecules in N_1 cells; since all N_1 molecules are identical, this number is unity, so that the entropy of the pure solvent is zero. Similarly the entropy of the pure solute is zero. Thus S in equation (3.22) represents the entropy of mixing which can be compared with the expressions in equations (3.19).

It is important to recognise the limitations of this treatment. Firstly, we have assumed that the solute and solvent molecules are of equal volume so that only one of each fits into a lattice cell. Secondly, we have assumed that all the configurations are of equal probability or that they have the same energy: this is consistent with the supposition that the energy of mixing is zero. Thirdly, we have only taken into account in equation (3.22) the entropy contingent on the mixing of the two sorts of molecule. We have not considered entropy changes originating from other sources such as differences in vibrational or rotational energy levels of the molecules in the pure state and in the solution. Nevertheless the agreement between equation (3.22) and the expression for the entropy of mixing of a two-component system implied by equations (3.19) suggests that all these assumptions are inherent in the definition of an ideal solution. Since similar factors must be taken into account in expressions for the entropy of mixing of all solutions, we see that an ideal solution represents a satisfactory baseline in terms of which all solutions may be considered.

3.6   Real solutions

Just as real gases do not necessarily obey the ideal gas laws, except at low pressures, real solutions do not behave as ideal solutions except under certain limiting circumstances.

To take such deviations from ideal behavior into account, we add another term into equation (3.17) and write

$$\mu_i = \mu_i^{\ominus} + RT \ln x_i + RT \ln f_i \tag{3.23}$$

where f_i is termed the activity coefficient of component i and is in general a function of the composition of the solution. Comparison of equation (3.23) with equation (3.14) shows that the product $x_i f_i$ is equal to the activity of the component.

Having introduced activity coefficients in this way we must consider the choice of standard states. For mixtures we have already indicated that the standard state is chosen to be that of the pure component so that $\mu_i^\ominus = G_i^*$. This implies that, as x_i tends to unity, μ_i must tend to $\mu_i^\ominus$. This in turn implies that $\ln f_i$ must simultaneously tend to zero or f_i tend to unity. Similar considerations apply to the chemical potential and activity coefficient of the solvent in a solution. The definition of the standard chemical potential of a solute in a solution is a little more subtle. We define the standard chemical potential of a solute to be the limit, as x_2 tends to zero, of the difference $\mu_2 - RT \ln x_2$. Each of the terms in this expression becomes infinite as x_2 tends to zero, but deeper considerations of the thermodynamics assures us that the difference remains finite. This being so, we see that the activity coefficient of the solute tends to unity as the concentration tends to zero. It should be noted that this is so whether the solute be macromolecular or not.

We now note that the activity coefficient of the solute in an ideal solution is, by definition, unity. Thus real and ideal solutions display the same dependence of their chemical potential on concentration at high dilution. In particular the relative chemical potential of the solvent in a real solution must tend to the same limit as implied by equation (3.20). That is to say, we may write for any solution

$$\mu_1 - \mu_1^\ominus = -RTV_1^* \rho_2 \left[\frac{1}{M_2} + B\rho_2 + C\rho_2^2 + \cdots \right] \tag{3.24}$$

where B and C are termed the second and third virial coefficients. The values of these will depend on the nature of the solute, but detailed statistical mechanical theories have been expounded which suggest that B depends on the nature of binary interactions between pairs of solute molecules, C depends on ternary interactions between triplets of molecules and so forth for the higher virial coefficients in the series.

The right-hand side of equation (3.23) contains three terms. The third of these, $RT \ln f_i$, is termed the *excess* component and is sometimes written as μ_i^E. It represents the free energy associated with interactions between the solute molecules at finite concentrations. For an ideal solution it is zero. The second term, $RT \ln x_i$, is termed the *cratic* component. It is entropic in origin and represents the free energy changes inherent in the mixing process. It is the same in ideal and non-ideal solutions alike. The first term, $\mu_i^\ominus$, is termed the *unitary* component. It includes all other free energy changes and is independent of the concentration. Thus it includes any free energy changes contingent on placing a solute molecule in isolation from its fellows in the solvent and so depends on solvent–solute interactions.

Finally we point out that it is sometimes desirable to express the chemical potential of a solute in terms of its mole concentration rather than its mole fraction. This leads to the necessity of redefining the standard chemical potential, the relative activity and the activity coefficient. We shall continue to use the symbol μ_i for the standard chemical potential and a_i for the activity since it will be apparent from the context which concentration scale is implied, but we shall denote the corresponding activity coefficient by y_i instead of f_i. It is readily shown that the activities of a component shared by two phases in equilibrium must be the same irrespective of which concentration scale is used. We also arrange that y_i tends to unity at infinite dilution like f_i, but it should be noted that y_i is in general not equal to unity at finite concentrations even for an ideal solution. Textbooks on thermodynamics provide expressions giving the relation between f_i and y_i but we shall make no use of these.

3.7   Colligative properties and osmotic pressure

We have seen above that the relative chemical potential of the solvent can be expressed as a power series in the concentration of the solute. At solute concentrations that are sufficiently low that the second and higher terms of this virial expansion can be ignored in comparison with the first we see that $\mu_1 - \mu_1^{\ominus}$ is proportional to ρ_2/M_2, which equals c_2. That is to say, it is proportional to the number of solute molecules present per unit volume. Measurable properties that are proportional to the relative chemical potential of the solvent are called *colligative* properties, and such properties include the relative lowering of the vapour pressure of the solvent, the elevation of the boiling point and the depression of the freezing point. For solutions of macromolecules, these properties are generally too small to measure; this is a direct consequence of their high molecular weight which implies that, for solutions containing a given mass concentration, ρ_2, there are several orders of magnitude fewer molecules than in the case of low-molecular-weight solutes. The only colligative property that is experimentally accessible for most solutions of macromolecules is the osmotic pressure of the solution.

Consider a vessel divided into two parts by a semipermeable membrane as illustrated in figure 3.2. The semipermeable membrane has the property of allowing solvent molecules to pass freely through it while preventing the passage of the much larger molecules of the macromolecular solute. It thus resembles a sieve but is otherwise inert.

The compartment on one side (labelled I) of the membrane is filled with pure solvent. That on the other side (labelled II) is filled with a solution of a macromolecule at a mass concentration, ρ_2. The solute has the effect of reducing the chemical potential of the solvent in which it is dissolved, which is thus lower than that of the solvent in compartment I. This implies that the system is not at equilibrium and that solvent will tend to flow through the membrane from I into II. Equilibrium can be restored, however, by increasing the pressure of the solution in compartment II so as to tend to cause solvent to flow back at the same rate as it flows in on account of the difference of

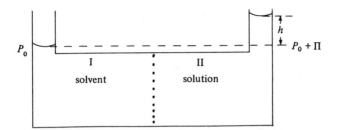

Figure 3.2. The diagram illustrates the arrangement in an osmotic-pressure experiment. Compartment I is filled with pure solvent. Compartment II is filled with solution at solute concentration ρ_2. The two compartments are separated by a semipermeable membrane indicated by the dotted lines. At equilibrium, the pressure in I is P_0 (atmospheric pressure) and that in II is $P_0 + \Pi$, where Π is the osmotic pressure, as indicated by the difference in levels of the solutions in the two compartments, with $\Pi = \rho g h$.

chemical potential. The condition of equilibrium can be found by noting that an increase in the pressure of a solution increases the chemical potential of the solvent. If the pressure is increased by an amount Π, the increase in the chemical potential of the solvent is given by

$$\int_{P_0}^{P_0 + \Pi} \left(\frac{\partial \mu_1}{\partial P}\right)_T dP$$

We therefore have at equilibrium, when the chemical potential of the solvent is the same on both sides of the membrane, the relation

$$\mu_1^I = \mu_1^\ominus = \mu_1 + \int_{P_0}^{P_0 + \Pi} \left(\frac{\partial \mu_1}{\partial P}\right)_T dP \tag{3.25}$$

where μ_1 is the chemical potential of the solvent in compartment II in the absence of the increased pressure and as such is equal to

$$\mu_1^\ominus - RTV_1^*\rho_2 \left[\frac{1}{M_2} + B\rho_2 + \cdots\right]$$

as obtained from equation (3.24).

It is readily shown that $(\partial \mu_1/\partial P)_T = \overline{V}_1$ and that for dilute solutions is essentially equal to V_1^*, which is almost independent of pressure. The integral in the right-hand side of equation (3.25) can therefore be evaluated to $V_1^*\Pi$. We thus obtain the result

$$\Pi = (\mu_1^\ominus - \mu_1)/V_1^* = RT\rho_2 \left[\frac{1}{M_2} + B\rho_2 + \cdots\right] \tag{3.26}$$

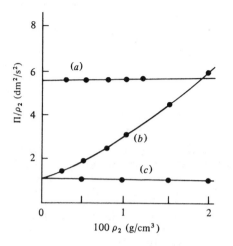

Figure 3.3. Plots of Π/ρ_2 versus ρ_2 for representative macromolecules. Π is the osmotic pressure and ρ_2 is the mass concentration of the macromolecular solute. The intercept of the curves at zero concentration gives the value of RT/M_2, where M_2 is the molar mass of the solute, and the slope at low concentrations is equal to RTB, where B is the second virial coefficient. Note that curve (b) is discernibly curved indicating that the third virial coefficient makes a significant contribution to the osmotic pressure. (a) The globular protein ovalbumin; (b) The synthetic polymer polyisobutylene in a good solvent (cyclohexane); (c) The same polyisobutylene fraction in a poor solvent (benzene).

The data for polyisobutylene are taken from P. J. Flory, *J. Am. Chem. Soc.*, **65**, 372, (1943). Those for ovalbumin from A. V. Guntelberg and K. Linderstrøm-Lang, *Comp. rend. trav. lab. Carlsberg, ser. chim.*, **27**, 1 (1949).

This excess pressure, Π, is known as the *osmotic pressure* of the solution. It can be measured, and is seen to be proportional at low concentrations of solute to ρ_2/M_2.

Measurements of the osmotic pressure can clearly be used to derive estimates of the molecular weight of the solute provided its mass concentration, ρ_2, is known. In practice it is necessary to measure the values of the osmotic pressure at several solute concentrations and to plot Π/ρ_2 against ρ_2 as illustrated in figure 3.3. The low concentration end of the plot can be extrapolated to zero solute concentration, and the extrapolated value of Π/ρ_2 is seen to be equal to RT/M_2.

Values of Π down to a value of about 10^2 Pa (or 1 cm of water) can be measured with reasonable precision. At room temperature with a 1% solution of solute, this osmotic pressure corresponds to a solute molecular weight of about 250 000, which represents an approximate upper limit to the range of molecular weights that can be measured accurately in this way. The lower limit is determined by the size of the pores of the membrane, which must be large enough to allow solvent molecules to pass, but small enough to prevent the passage of the macromolecules.

Various devices are commercially available for measuring osmotic pressures. They generally work by monitoring the direction and magnitude of the flow of solvent through the membrane and arranging for the pressure of the solvent in compartment I to be reduced till this flow is zero. The pressure difference between the two compartments is then measured manometrically. It is important for the temperature to be closely controlled in order to avoid random fluctuations that disturb the equilibrium.

It should be noted that, if the second virial coefficient is large, it may be necessary to measure the osmotic pressure at very low concentrations in order to achieve a reliable extrapolation to zero concentration. This is especially so if the macromolecule exhibits a tendency to aggregate.

It is interesting to examine the value of the osmotic pressure if the macromolecular solute is paucidisperse or polydisperse. If we assume that the concentration, ρ_2, that is measured and employed in the calculation is the total mass concentration of all macromolecular solutes it is clear that $\rho_2 = \Sigma_i c_i M_i$, where c_i is the concentration of the species with molar mass M_i. The total concentration of all species is clearly $\Sigma_i c_i$. This is equal to the ratio ρ_2/M_{av}, where M_{av} is the average molar mass obtained by extrapolation to infinite dilution. We thus see that $M_{av} = \Sigma_i c_i M_i/\Sigma_i c_i$ and as such is equal to the number average as defined by equation (1.1).

3.8   The origin of non-ideality

The cause of departures from ideality by solutions at finite concentrations is to be traced through the excess chemical potential or the virial coefficients to interactions between the solute molecules. As we shall see below, solutions of macromolecules are intrinsically non-ideal on account of their large size compared to the size of the solvent molecules. First let us note, however, that the excess chemical potential of a component can be split into two terms, in the manner of equation (3.16), into an excess partial molar enthalpy, $\overline{H}_i^E$ and a term in the excess molar entropy $\overline{S}_i^E$ such that $\mu_i^E = \overline{H}_i^E - T\overline{S}_i^E$. We shall consider both of these components separately for the case of mixtures of flexible linear polymers in a low-molecular-weight solvent, that is to say, concentrated solutions. We shall also consider the case of dilute solutions of macromolecules, both flexible polymers and rigid rod-like and globular macromolecules.

In order to investigate these excess functions we shall again invoke a lattice model. As before, this will consist of a large number of cells into each of which may be placed either a solvent molecule or a *polymer segment*. The latter is to be conceived of as a monomer unit or a small number of monomer units equal in volume to a solvent molecule. In the case of polystyrene dissolved in toluene, each monomer is approximately the same size as a solvent molecule so that a segment is equivalent to a monomer unit. For other polymers this is not necessarily the case.

We shall first consider *athermic* polymer solutions, in which the energy of interaction between two segments in neighbouring cells is the same as the energy of interaction between a segment next to a solvent molecule or between two adjacent solvent molecules.

3.9   The entropy of mixing of athermic polymer solutions

The fundamental difference between an ideal solution of small molecules and an athermic polymer solution is that, in the latter, the segments that can occupy lattice sites are connected, whereas the solute molecules in an ideal solution are not. We must therefore investigate what difference this connectedness makes to the thermodynamic properties of the solution. To do this we shall count the number of ways in which polymer and solvent molecules can be placed in the lattice to obtain an estimate of the number of possible and distinguishable configurations, which will in turn lead to an estimate of the entropy of the solution. For an athermic solution, all these possible configurations will be equally probable.

Let us suppose that the solution contains N_1 solvent molecules and N_2 polymer molecules, each consisting of σ segments. The lattice will therefore contain $N_1 + \sigma N_2 = N_0$ cells. We shall also suppose that the coordination number of the lattice is z; that is to say each cell is surrounded by z others.

Let us suppose that N_i polymer chains have already been placed in the lattice and consider the number of ways of placing the $(N_i + 1)$th. These first N_i chains occupy σN_i cells which we shall suppose represents a fraction, r_i, of the total number, N_0. There are thus $N_0 - \sigma N_i = (1 - r_i) N_0$ vacant cells awaiting the next chain. The first segment of the next chain can be placed in any one of these vacant cells. The second segment could then be placed in any vacant cell that is adjacent to the first. The probability that one of the z adjacent cells is vacant is $(1 - r_i)$, assuming that the cells are occupied at random. Thus, on average, the second segment can be placed in $z(1 - r_i)$ ways. Since the third segment cannot occupy the same cell as the first, there are on average only $(z - 1)(1 - r_i)$ ways in which it can be placed, and this is also the number of ways in which each of the remaining segments can be placed. A more exact treatment would take into account the fact that r_i should be increased each time a new segment is put into the lattice, but to ignore this is to make but a small approximation in comparison with others inherent in this treatment. We see then that the total number of ways in which the σ segments of the $(N_i + 1)$th chain can be placed in the lattice is

$N_0(1 - r_i)^\sigma z(z - 1)^{\sigma-2}$. We shall simplify this expression by replacing the solitary factor z by $z - 1$ and by noting that $N_0(1 - r_i) = N_0 - \sigma N_i$.

We then see that the total number of ways of placing all N_2 chains into the lattice is the product of N_2 terms of this kind. As discussed in section 3.5 we must divide the result by $N_2!$ to take into account the indistinguishability of the N_2 solute molecules. Again as discussed in section 3.5, the remaining cells can be filled with solvent molecules in only one way. We thus see that the number of configurations of the mixture is

$$\Omega = \frac{1}{N_2!} \left(\frac{z - 1}{N_0} \right)^{N_2(\sigma-1)} \prod_{N_i=0}^{N_2-1} (N_0 - \sigma N_i)^\sigma \qquad (3.27)$$

We need to evaluate $k \ln \Omega$. When we take the logarithm of Ω we obtain, among other terms, a sum of terms of the form $\sigma \ln (N_0 - \sigma N_i)$ with N_i taking the values 0 to $N_2 - 1$. This sum may be replaced by an integral since N_2 is very large. We may also replace the logarithm of the factorial by Stirling's approximation. Proceeding in this way we can evaluate an expression for the entropy associated with the solution. If we put N_1 and N_2 in turn equal to zero we obtain the entropy associated with the pure solute and the pure solvent respectively. If these are subtracted from the entropy associated with the solution we finally obtain an expression for the entropy of mixing, ΔS_m. This may be written in the pleasingly simple form

$$\Delta S_m = -R(n_1 \ln \phi_1 + n_2 \ln \phi_2) \qquad (3.28)$$

where ϕ_1 and ϕ_2 are the fractions of the number of cells occupied by solvent and solute in the solution. They are usually referred to as their *volume fractions* on the basis of the fact that the cells are of equal volume. They are thus defined by

$$\phi_1 = \frac{N_1}{N_1 + \sigma N_2} \quad \text{and} \quad \phi_2 = \frac{\sigma N_2}{N_1 + \sigma N_2} \qquad (3.29)$$

This remarkably simple result is seen to reduce to the entropy of mixing of an ideal solution as given by equation (3.19) if σ is set equal to unity. It also implies that solutions of flexible polymers are inherently non-ideal. An interesting feature of the result is that it is independent of z, the coordination number of the lattice. This suggests that the result has a greater generality than is implied by the use of the artifice of the lattice; other treatments of the problem confirm this conclusion.

Apart from approximations inherent in the use of a lattice and those introduced in the derivation of equation (3.28), it suffers from two grave limitations. The first is that it only applies to athermic solutions, in which every possible configuration is equally probable; this limitation will be discussed further in the next section. The second is due to the assumption that the probability, $1 - r_i$, that a cell is vacant is independent of its position in the lattice. In a very dilute solution, the mean segment density in the vicinity of a polymer molecule will be much greater than in a region of the lattice well separated from any such molecule. Thus dilute solutions of flexible polymers require a separate treatment as given below, and equation (3.29) only applies to concentrated solutions or mixtures.

Finally, it should be emphasised that equation (3.24) is based on the simplest possible treatment. Since the theory was first proposed by Flory and Higgins, other workers have elaborated on it to improve on the assumptions involved.

3.10   The energy of mixing of polymer solutions

In this section we consider the necessary extensions of the lattice theory of solutions of flexible polymers when they are no longer assumed to be athermal. To do this we must consider interactions between nearest neighbours in the lattice.

There are three sorts of nearest neighbour in the lattice, and we shall suppose that each is associated with a characteristic interaction energy. Thus we suppose that V_{11} is the energy of interaction of two solvent molecules in adjacent cells; that V_{12} is the energy for a polymer segment adjacent to a solvent molecule; and that V_{22} is the energy for two adjacent polymer segments that are not directly connected in the same chain. We shall ignore interactions between adjacent segments in the same chain since there are the same number of these in the solution as in the pure polymer solute.

We shall assume that the mixing is random, even though we would expect in reality that interactions with the lower energies would be statistically favoured. This assumption, known as the Bragg-Williams approximation, can be removed, but only at the expense of a great complication of the theory. We shall also assume that only interactions between nearest neighbours in the lattice are of significance. This is reasonable if the interaction energy falls off according to a large inverse power of the separation distance.

We now recollect that each cell in the lattice has z nearest neighbours. A fraction, ϕ_2, of these will contain polymer segments on average, and a fraction, $1 - \phi_2$, will contain solvent molecules. Thus each solvent molecule is surrounded, on average, by $z\phi_2$ segments and $z(1 - \phi_2)$ other solvent molecules. Its energy of interaction with its neighbours is then given by $z\phi_2 V_{12} + z(1 - \phi_2) V_{11}$. Similarly the energy of interaction of a polymer segment with its neighbours is $z\phi_2 V_{22} + z(1 - \phi_2) V_{12}$. Strictly speaking z in this latter expression should be replaced by $z - 2$ to take account of the fact that each segment has two neighbouring segments which are its neighbours in the same chain (except for the rare segments at the end of the chain). Using z rather than $z - 2$ implies that we are tacitly ignoring the fact that the segments are connected, but the effect of this approximation is small compared with other approximations inherent in this treatment.

We may now sum up the interaction energies of all $N_1 (= N_0 \phi_1)$ solvent molecules and all $\sigma N_2 (= N_0 \phi_2)$ segments. This sum should then be divided by two to give the total interactions energy in the solution since each interaction would otherwise be counted twice. If we put N_1 and N_2 in turn equal to zero we then obtain the interaction energy associated with the pure solute and pure solvent respectively. These may then be subtracted from the energy associated with the solvent to obtain the energy of mixing, ΔU_m, which may be written in the form

$$\Delta U_m = -\tfrac{1}{2} z N_0 \phi_1 \phi_2 (V_{11} + V_{22} - 2V_{12}) = RT \phi_2 n_1 \chi_1 \tag{3.30}$$

Here we have defined the interaction parameter, χ_1. If $\chi_1 < 0$, solvent-segment interactions are favoured, and the solvent may be described as a good solvent. Conversely, if $\chi_1 > 0$, solvent-segment interactions are disfavoured, and the solvent is a poor solvent. If $\chi_1 = 0$, the solution is athermic.

The formalism of the lattice model does not readily accommodate the possibility of there being a volume change of mixing. Such changes in volume are likely to occur in the case of polymer molecules bearing polar side groups dissolved in polar solvent such as water. If, however, we bear in mind this limitation of the theory we may equate the energy of mixing to the enthalpy of mixing, ΔH_m.

We may then combine equations (3.28) and (3.30) to obtain an expression for the Gibbs free energy of mixing, ΔG_m

$$\Delta G_m = \Delta H_m - T\Delta S_m = RT(n_1 \ln \phi_1 + n_2 \ln \phi_2 + \chi_1 \phi_2 n_1) \tag{3.31}$$

If we differentiate this expression with respect to n_1 and with respect to n_2 we obtain expressions for the relative chemical potentials of the solvent and solute respectively:

$$\mu_1 - \mu_1^{\ominus} = RT \left[\ln (1 - \phi_2) + \phi_2 \left(1 - \frac{1}{\sigma} \right) + \chi_1 \phi_2^2 \right] \tag{3.32}$$

$$\mu_2 - \mu_2^{\ominus} = RT [\ln \phi_2 + (1 - \phi_2)(1 - \sigma) + \chi_1 \sigma (1 - \phi_2)^2] \tag{3.33}$$

Expansion of the logarithmic term in equation (3.32) then leads to the virial expansion of the relative chemical potential of the solvent:

$$\mu_1 - \mu_1^{\ominus} = -RTV_1^* \rho_2 \left[\frac{1}{M_2} + \left(\frac{1}{2} - \chi_1 \right) \left(\frac{V_2^*}{M_2} \right)^2 \frac{\rho_2}{V_1^*} + \cdots \right]. \tag{3.34}$$

where we have equated the volume of N_A lattice cells to V_1^*, the molar volume of the pure solvent and that of σN_A cells to the molar volume of the pure polymer V_2^*. N_A is Avogadro's constant. If equation (3.34) is compared with the expression given by equation (3.20) for an ideal solution we may discern that the excess chemical potential of the solvent, μ_1^E, is given by

$$\mu_1^E = -RT\rho_2^2 \left[\frac{1}{2M_2^2}(V_2^{*2} - V_1^{*2}) - \left(\frac{V_2^*}{M_2}\right)^2 \chi_1 + \cdots \right] \tag{3.35}$$

The first term in the brackets represents the excess partial molar entropy contribution originating from the disparity between the volumes of the polymer and solvent molecules; it is zero if they are of equal volume ($V_2^* = V_1^*$). The second term represents the excess partial molar enthalpy originating from differential interactions between segments and solvent molecules; it is zero if $\chi_1 = 0$.

These results are based on numerous assumptions which we have noted as we have proceeded in their derivation. We must therefore not expect that they will afford an accurate representation of the behaviour of real polymer solutions. We emphasise that they apply only to solutions of flexible polymers in concentrated solutions. The consequences of some of the assumptions will be less serious if the interactions between two polymer segments, between a polymer segment and a solvent molecule and between two solvent molecules are associated with comparable interaction energies. Such is the case for a non-polar polymer, such as polymethylene, dissolved in a non-polar solvent such as dodecanol.

3.11   The interaction parameter

The dimensionless interaction parameter, χ_1, has been defined in equation (3.30) such that $kT\chi_1/z = -\frac{1}{2}(V_{11} + V_{22} - 2V_{12})$. This latter parameter represents the energy change involved in the formation of one segment–segment contact in the solution. As such it does not take into account any entropy change contingent on this 'reaction', although such entropy changes are normally to be expected.

This restriction on the validity of equation (3.34) can be lifted by replacing the energy term $(\frac{1}{2} - \chi_1)RT$ which occurs in this equation by a term of the form $-(\Delta H_{int} - T\Delta S_{int})$, where ΔH_{int} represents the energy change involved in the 'reaction' and ΔS_{int} represents the entropy change together with the small term, $\frac{1}{2}R$, which occurs in $(\frac{1}{2} - \chi_1)R$ and which it is convenient to absorb into ΔS_{int}.

It is usual to define two further parameters: Θ as $\Delta H_{int}/\Delta S_{int}$ and Ψ as $\Delta S_{int}/R$. We then see that $(\frac{1}{2} - \chi_1)$ is to be replaced by $\Psi(1 - \Theta/T)$. Ψ is clearly dimensionless, but Θ has the dimensions of temperature.

If the temperature is equal to Θ, we see that the second virial coefficient expressed in this new form becomes zero. Such a temperature is referred to as the *theta temperature*, or more generally, if the properties of the solvent are such that $\Theta = T$, we may refer to a *theta solvent*. At temperatures below the theta temperature, the solvent becomes a poor solvent and, at temperatures above the theta temperature, a good solvent.

We shall see in later sections that under theta conditions the second virial coefficient of even dilute polymer solutions becomes zero and, to anticipate arguments presented in the next chapter, the configurational statistics of flexible polymers take a particularly simple form.

It is found in practice that a temperature or a solvent can indeed be found at which the second virial coefficient is zero. This can be done by measuring the osmotic pressure as a function of concentration at a series of different temperatures. At the theta temperature, Π/ρ_2 is independent of concentration.

Finally it should be noted that the replacement of $(\frac{1}{2} - \chi_1)$ by $\Psi(1 - \Theta/T)$ obscures the distinction between the excess partial molar entropy and enthalpy that we have discerned in equation (3.35).

3.12 Phase equilibria in poor solvents

Figure 3.4 shows plots of the relative chemical potential of the solvent, $-(\mu_1 - \mu_1^{\ominus})/RT$, versus the volume fraction of the polymer solute, ϕ_2. These are based on equation (3.32) and assume various values of the interaction parameter, χ_1. Examination of these plots reveals that there is a critical value of χ_1 at which the curves begin to show a point of inflection. For values of χ_1 above this critical value there is a range of values of the ordinate in which there are two different corresponding values of ϕ_2. Within this range it is readily seen that, for each value of χ_1, there are two different values of ϕ_2 at which the chemical potentials of both the solvent and the solute are the same. This implies that solutions with concentrations defined by these two values of ϕ_2 can be in thermodynamic equilibrium. Moreover, it implies that a solution with an intermediate value of ϕ_2 will spontaneously separate out into two phases with these two concentrations; this will occur with a concomitant decrease in the free energy.

Such phenomena are in fact observed with real solutions of flexible polymers for values of χ_1 above a critical value. Thus, if χ_1 is increased either by decreasing the temperature or by altering the nature or composition of the solvent, phase separation preceded by the onset of an opalescence is commonly observed.

The critical point is located by the conditions for a point of inflection that $\partial \mu_1/\partial \phi_2 = \partial^2 \mu_1/\partial \phi_2^2 = 0$. Application of these conditions to equation (3.32) leads to the following expressions for the critical value of the interaction parameter, χ_{1c}, and the value of ϕ_2 at the point of inflection, ϕ_{2c}:

$$\chi_{1c} = (1 + \sqrt{\sigma})^2/2\sigma \tag{3.36}$$

$$\phi_{2c} = 1/(1 + \sqrt{\sigma}) \tag{3.37}$$

If the molecular weight of the polymer is very large these expressions approximate to $\chi_{1c} = \frac{1}{2}$ and $\phi_{2c} = 1/\sqrt{\sigma}$.

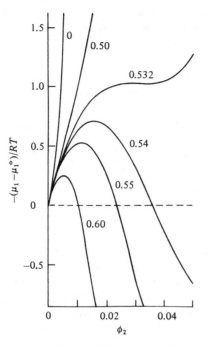

Figure 3.4. Plots, based on equation (3.32), of $(\mu_1 - \mu_1^{\ominus})/RT$ versus ϕ_2. The quantity $\mu_1 - \mu_1^{\ominus}$ is the relative chemical potential of the solvent and ϕ_2 the volume fraction of the macromolecular solute. A value of 1000 has been assumed for σ, the number of segments in the polymer. The different curves correspond to various values of the interaction parameter, χ_1, as indicated in the diagram.

If $(\frac{1}{2} - \chi_1)$ is replaced by $\Psi(1 - \Theta/T)$, we see that there is a critical temperature, T_c, corresponding to the critical value of χ_1. It is readily shown that

$$\Psi(\Theta/T_c - 1) = 1/2\sigma + 1/\sqrt{\sigma} \qquad (3.38)$$

This critical temperature may be determined by observing the onset of opalescence as a solution of the polymer is cooled. If T_c is thus determined for a series of fractionated polymers with different degrees of polymerisation, σ, a plot of $1/T_c$ versus $1/2\sigma + 1/\sqrt{\sigma}$ should be linear with a slope of $\Psi\Theta$ and an intercept at infinite σ of Θ. Such plots therefore should yield values of Ψ and Θ which should depend only on the nature of the polymer and the solvent and not on the degree of polymerisation.

In mixtures of two or more polymers differing in molecular weight or in the nature of their monomer units, the solutions which separate out below the critical temperature will generally contain different proportions of the polymeric components. The phenomenon of phase separation may therefore be used to fractionate polymer preparations on the basis of molecular weight or other properties.

3.13 Experimental tests of the theory of concentrated polymer solutions

The relations developed in the previous sections have been submitted to numerous experimental tests using a variety of flexible polymers in a variety of solvents. In general terms the results have been in qualitative but not quantitative agreement with the theory. This is not unexpected considering the extreme simplicity of the model on which the theory is based and the approximations entailed in its development.

In particular, estimates of χ_1 or of ΔH_{int} and ΔS_{int} are frequently dependent on the concentration ϕ_2; effects of this nature are more frequently encountered in systems in which either the polymer or the solvent is polar in nature.

A useful test of the validity of equation (3.32) is afforded by noting that it implies that $\mu_1 - \mu_1^{\ominus} - \ln \phi_1 - (1 - V_1^*/V_2^*) \phi_2$ is proportional to ϕ_2^2. The relative chemical potential $\mu_1 - \mu_1^{\ominus}$ may be determined from osmotic-pressure measurements so that both of these expressions may be evaluated for a series of different concentrations. If the first is plotted against the second, a straight-line plot should result with a slope equal to χ_1. Such plots are indeed frequently found to be linear (except at low concentrations when the theory is not applicable), but deviations from linearity are encountered in some systems, particularly in poor solvents.

3.14 Dilute solutions

As we have reiterated, the lattice theory of flexible-polymer solutions discussed above is invalid for dilute solutions, in which the domains inhabited by the individual polymer molecules are far apart. To cope with dilute solutions, a different approach is required and for this reason we invoke the concept of the *excluded volume.*

The excluded volume of a solute molecule is the volume that is not available (because of exclusion forces or for other reasons) to the centres of mass of other similar solute molecules. As an example we may consider the excluded volume of a spherical particle of radius, R. This excluded volume is equal to $\frac{4}{3}\pi(2R)^3$ since the centres of two such spheres may not approach nearer than a distance $2R$.

We next note that the entropy associated with a single solute molecule in a solution is dependent on the volume available to it. Thus the more solvent molecules there are in a given volume of solution, the less volume there is available to any one of them, and hence the lower is the entropy associated with each.

In developing a theory of dilute solutions along these lines two problems present themselves. The first is to relate the free energy of the solution to the concentration and the excluded volume of the solute molecules. The second is to relate the excluded volume to the molecular properties of the large molecules.

3.15   The free energy of mixing of dilute solutions

First recollect that the standard state of the solute in a dilute solution is taken as a state of infinite dilution. In this standard state the solute molecules are fully solvated and otherwise in a state of equilibrium with the surrounding solvent. We may imagine the formation of a dilute solution by a process of mixing such fully solvated solute molecules with solvent. We shall assume for the present that such fully solvated solute molecules do not interact energetically with each other. It then follows that the process of mixing is accompanied by no change in the enthalpy and that the free energy of mixing is wholly entropic. We may thus write

$$\mu_1 - \mu_1^\ominus = -T(\bar{S}_1 - S_1^*) \tag{3.39}$$

where S_1^* is the molar entropy of the pure solvent and $\bar{S}_1$ is the partial molar entropy of the solvent in the solution.

In order to evaluate this entropy difference we shall also assume, as is reasonable, that the entropy and energy associated with rotational, vibrational (etc.) motion of the component molecules is not changed by the mixing process.

Consider now a solution of volume, V, containing N_2 solute molecules, each of excluded volume, u. We proceed to build up the solution by adding the solute molecules, one at a time, to the solvent. The volume available to the first molecule is the total volume of the solution, V, and we assume that the number of ways it may be placed in this volume is proportional to V. Thus the first molecule may be placed in KV ways, where K is a constant of proportionality. Having inserted the first molecule, the volume available to the second is $V - u$ so that it may be placed in $K(V - u)$ ways. Similarly, the third molecule may be placed in $K(V - 2u)$ ways and so forth for the remainder. We thus see that the total number of distinguishable ways that all N_2 indistinguishable solute molecules can be placed is

$$\Omega = \frac{1}{N_2!} \prod_{j=0}^{N_2-1} K(V - ju) \tag{3.40}$$

where the total volume of the solution may be replaced by $n_1 \bar{V}_1 + n_2 \bar{V}_2$ as indicated by equation (3.4). The entropy associated with the solution on account of the possible ways of placing the solute molecules is then given by $k \ln \Omega$. This may be evaluated from equation (3.40) by invoking Stirling's approximation for the logarithm of the factorial and by expanding the logarithmic terms.

In this result we may put n_1 and n_2 in turn equal to zero to obtain the entropy associated with the solute and solvent in their standard states. Subtracting these from the entropy associated with the solution gives the entropy of mixing, ΔS_m.

Finally we differentiate the expression so obtained for the entropy of mixing with respect to n_1 to obtain, according to equation (3.39), the relative chemical potential of the solvent. This final result may be expressed

$$\mu_1 - \mu_1^\ominus = -RTV_1^* \rho_2 \left[\frac{1}{M_2} + \frac{N_A u}{2M_2^2} \rho_2 \right] \tag{3.41}$$

where N_A is Avogadro's constant and we have expressed the concentration of the solute as a weight concentration, ρ_2. We have also replaced $\overline{V}_1$ by V_1^*, which is a valid approximation for dilute solutions.

It is important to note that equation (3.41) is only valid for dilute solutions. This is because, in the first place, we have tacitly ignored the possibility of ternary and higher encounters between solute molecules. These entail that the volume unavailable to any given solute molecule is not the sum of the excluded volumes of the remainder, but somewhat less. A proper account of these would require terms in higher powers of the concentration, ρ_2, than appear in equation (3.41). In the second place we have assumed that there is no energy change contingent on solute-solute interactions. This will only be appropriate if such encounters are rare.

Finally we note that if u, the excluded volume of a solute molecule, is put equal to the volume of a solvent molecule, V_1^*/N_A, the second term in equation (3.41) reduces to the second term in the virial expansion for an ideal solution as given by equation (3.20).

The next problem is to consider the excluded volume of various categories of macromolecule.

3.16   The excluded volume of rigid molecules

We have indicated above that the excluded volume of a spherical particle is eight times its physical volume, and that the factor of eight arises because the centres of two spheres cannot approach nearer than a distance of twice their common radius. The excluded volume of asymmetric particles cannot be calculated so easily. This is because the distance between their centres of mass when they are in contact depends on their orientation. Nevertheless it has been calculated that the excluded volume of a rod of diameter, d, and length, L, is $\frac{1}{2}dL^2$.

Of course no real molecule presents a surface as smooth as that of a sphere or a rod, except as a rough approximation, so that accurate estimations of the excluded volumes of rigid molecules such as globular proteins or helical rod-like macromolecules is out of the question, and rough approximations must suffice. It should also be borne in mind that it is the excluded volume of the fully solvated particle that is of interest. If the particle is extensively solvated, the layer of bound solvent will effectively increase the excluded volume. It is partly for these reasons that measurements of the second virial coefficient (from osmotic-pressure measurements or light-scattering data) cannot, in general, be used to provide reliable estimates of the dimensions of rigid macromolecules.

Another reason for this exists if the macromolecule exhibits a tendency to aggregate. The presence of aggregates effectively decreases the number of particles present and so decreases the osmotic pressure and other colligative properties. Such effects may be manifest in negative second virial coefficients.

3.17   The excluded volume of flexible polymers

An isolated linear polymer molecule in solution continuously changes between a myriad of possible conformations as a result of random bombardment by solvent molecules. In the vicinity of its centre of mass, a small volume element will contain a fluctuating number of polymer segments. Although the density of segments in this element is continually changing, it has a well-defined average value. This time-average density of segments will correspond to a volume fraction of segments that is generally small in comparison with unity. It might therefore be thought that there is plenty of free space within the domain inhabited by one polymer molecule to allow a second to interpenetrate it. This is of course what happens in concentrated solutions of flexible polymers, but we have seen that this is accompanied by a pronounced decrease in the entropy, which leads to pronounced departures from ideality. This decrease in entropy which occurs when the two domains interpenetrate renders such interpenetration thermodynamically unfavourable and is the origin of the excluded volume of flexible-polymer chains. The decrease in entropy is only counteracted in poor solvents in which intramolecular contacts between polymer segments are favoured. This implies, as we shall see, that the excluded volume of flexible polymers depends on the temperature and the nature of the solvent.

In order to put these ideas on a more quantitative footing we shall discuss the original theory of Flory and Krigbaum, albeit in a simplified form.

We shall suppose that the domain of a polymer molecule is spherical with a uniform distribution of segments within it. This latter supposition, that the distribution of segments is uniform throughout the domain, represents a severe oversimplification as we shall see in chapter 4. Nevertheless, it turns out that the precise details of the distribution do not affect the final result greatly.

We first consider the free energy of mixing the segments and solvent within this domain. This is given by equation (3.31). Since there is only one polymer molecule within the domain, $n_2 = 1/N_A$, so that the second term in the expression for ΔG_m is very small and can be ignored. If the volume of the domain is defined to be V_d/N_A we thus see that the free energy of mixing per unit volume of the domain is given by $\Delta G_m^d(\phi_2)$ in

$$\Delta G_m^d(\phi_2) = \frac{N_1 RT}{V_d} \left[\ln (1 - \phi_2) + \chi_1 \phi_2\right] \tag{3.42}$$

where N_1 is the total number of solvent molecules in the domain and ϕ_2 is the volume fraction that the segments occupy in the domain. If we assume that the volume of each solvent molecule is V_1^*/N_A we can readily perceive that $N_1 = (1 - \phi_2) V_d/V_1^*$. We thus see that ΔG_m^d may be expressed as a function of ϕ_2 as the only variable, and we emphasise this by writing it as $\Delta G_m^d(\phi_2)$ in equation (3.42).

Let us now suppose that the domains of two polymer molecules approach one another from a large distance till they partially overlap. Let us also suppose that the fraction of the volume of either domain that is involved in the overlap is F. We thus see that a region of space of volume FV_d/N_A in which the volume fraction of segments is $2\phi_2$ has appeared and two regions of space each of the same volume, FV_d/N_A, and in which the segment volume fraction is ϕ_2, have disappeared. The appearance and disappearance of these regions is associated with free energy changes, and it is clear that the net free energy change is $[\Delta G_m^d(2\phi_2) - 2\Delta G_m^d(\phi_2)] FV_d/N_A$. This expression may be evaluated with the aid of equation (3.42) and simplified by expanding the logarithmic term and retaining only terms in ϕ_2 and ϕ_2^2. The final result, which gives the free energy change associated with the overlapping of two domains, is given by ΔG in

$$\Delta G = 2kTF(\tfrac{1}{2} - \chi_1) V_2^{*2}/V_d V_1^* \tag{3.43}$$

In this expression we have eliminated ϕ_2 by noting that it is, to a first approximation, equal to V_2^*/V_d, where V_2^* is the molar volume of the polymer.

Let us now suppose that the radius of the spherical domains is R_d so that $V_d/N_A = \tfrac{4}{3}\pi R_d^3$, and also suppose that they overlap in such a way that the distance between their centres is $2R_d y$ ($0 < y < 1$). It is easy to show then that the volume of the lens-shaped region that they have in common is $\tfrac{1}{2} V_d(y^3 - 3y + 2)/N_A$ so that the fraction, F, is

$$F = \tfrac{1}{2}(y^3 - 3y + 2) \tag{3.44}$$

We shall also find it convenient to define a constant, X, which depends only on the properties of the solvent and polymer by

$$X = (\tfrac{1}{2} - \chi_1) V_2^{*2}/V_d V_1^* \tag{3.45}$$

so that $\Delta G = 2kTFX$.

For the reasons that have been discussed in section 3.11, the factor $(\tfrac{1}{2} - \chi_1)$ in equation (3.45) may be replaced by $\Psi(1 - \Theta/T)$. We thus see that the free energy of intersection of two domains is zero at the theta temperature or in a theta solvent. Under these conditions, there is no energy barrier to even complete interpenetration of the two domains, and each behaves as if the other were not there. This implies that the excluded volume of the flexible polymer is zero! At temperatures below the theta point or in poor solvents the free energy of intersection is negative, which implies that under these circumstances the polymer molecules will tend to aggregate. Conversely, in good solvents the positive free energy change will render such intersections improbable, and there will be a finite effective excluded volume to which we must now turn our attention.

Suppose that a certain region of the solution contains only two polymer molecules, occupying domains labelled I and II. Suppose that I is fixed in space and that II is free to move. We further suppose that the probability of finding the centre of II in a small volume element, dV, is proportional to dV. If the element is situated so that the two domains do not overlap, this probability, $P(\infty)\,dV$ is a constant, independent of the position of the element. If, on the other hand, it is so situated that the domains do overlap, so that the distance between their centres is $2R_d y$, the probability, $P(y)\,dV$, depends on the free energy change contingent on the overlap, ΔG, which depends on y. Since ΔG is the free energy change that occurs when a non-overlapping situation is turned into an overlapping one, the Boltzmann distribution law tells us that the ratio $P(y)/P(\infty)$ is equal to $e^{-\Delta G/kT}$ or e^{-2FX}. Thus this ratio falls from a value of unity when $y > 1$ to a small value as y decreases to zero, in the manner shown in figure 3.5. Let us define f_y to be this ratio, so that $f_y = e^{-2FX}$.

Now the total volume such that the centres of the two domains are between distances $2R_d y$ and $2R_d(y + dy)$ apart is that of a spherical shell of radius $2R_d y$ and thickness $d(2R_d y)$. This volume is $32\pi R_d^3 y^2\,dy$ so that the probability of finding II in such a volume is $32\pi R_d^3 P(y)\,y^2\,dy$. Thus the total probability of finding it at a distance from I, such that overlap at some value of y less than 1 occurs, is the integral $32\pi R_d^3 \int_0^1 P(y)\,y^2\,dy$. It is clear, however, that total volume in which II must be situated for overlap to occur is that of a sphere of radius $2R_d$ and volume $\tfrac{32}{3}\pi R_d^3 = 8V_d/N_A$. This volume we will consider to consist of two regions: a region of volume u, the excluded volume in which the probability of finding the centre of II is zero, and the rest of volume $(32\pi R_d^3/3) - u$, in which the probability of finding II is

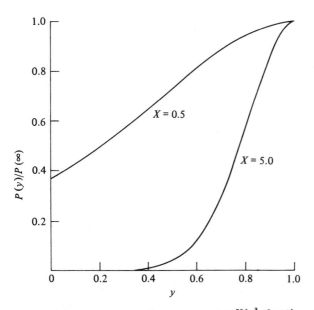

Figure 3.5. Plots of $P(y)/P(\infty) = e^{-X(y^3 - 3y + 2)}$ versus y. $2ry$ is the distance between the centres of two domains and X is defined by equation (3.45). Two curves are shown, one for $X = 0.5$, which corresponds to a poor solvent, and the other for $X = 5$, which corresponds to a good solvent.

$[(32\pi R_d^3/3) - u] P(\infty)$. This latter expression also represents the total probability of finding the centre of II within a distance $2R_d$ from the centre of I. We may thus equate the probability $32\pi R_d^3 \int_0^1 P(y) y^2 \, dy$ to $[(32\pi R_d^3/3) - u] P(\infty)$ and so find that

$$u = \frac{24V_d}{N_A} \int_0^1 (1 - f_y) y^2 \, dy \tag{3.46}$$

To proceed further we substitute e^{-2FX} for f_y and the expression given in equation (3.44) for F. The resulting integral cannot be evaluated explicitly, but after an integration by parts we find

$$u = 2V_d X h(X)/N_A \tag{3.47}$$

where the function $h(X)$ is

$$h(X) = -12 \int_0^1 y^3 (y^2 - 1) e^{-X(y^3 - 3y + 2)} \, dy \tag{3.48}$$

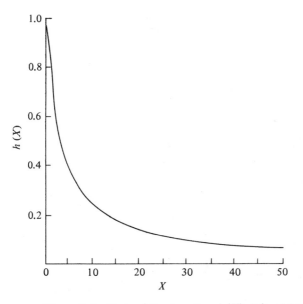

Figure 3.6. Plots of the function $h(X)$ defined by equation (3.48) versus X.

A plot of $h(X)$ versus X is shown in figure 3.6, and it is easy to show that, for $X = 0$ (as occurs under theta conditions), $h(X) = 1$ and that, as X tends to infinity, $h(X)$ tends to zero.

The value of X (given by equation (3.45)) increases with the 'goodness' of the solvent. Thus in ideal solvents which give theta conditions, $X = 0$, $u = 0$ and the second virial coefficient is zero, just as for concentrated solutions of flexible polymers. Similarly, if X is small, the excluded volume is approximately $2V_d X/N_A$, and the second virial coefficient is the same as the value given for concentrated solutions by equation (3.34). It is only in good solvents giving large values of X that the behaviours of dilute and concentrated solutions of flexible polymers diverge. Under these circumstances the excluded volume becomes $8V_d/N_A$, which implies that the domains behave like hard impenetrable spheres.

As we have mentioned above, the assumption that the domains are of uniform segment density is highly artificial. Nevertheless, the more realistic assumption that the segments are distributed in a Gaussian manner about the centre of mass leads to closely similar results. If the radius of the domain, R_d (which has so far not been related to the properties of the polymer), is equated to $(4\pi/3)^{1/6}$ times the radius of gyration of the distribution (as defined in chapter 4), the results for the two models become identical, except that the function $h(X)$ must be replaced by another function of X which has the same limit as $h(X)$ when $X = 0$.

In poor solvents, in which χ_1 is small, the virial coefficient is proportional to V_2^{*2}/M_2^2 and so is independent of the molecular weight. In good solvents we find that B is proportional to V_d/M_2^2. In chapter 4 we shall see that the radius of gyration is proportional to $M_2^{0.6}$ in good solvents, so that B is proportional to $M_2^{-0.2}$.

The validity of this theory of dilute solutions depends on the theory of concentrated solutions, since equation (3.31) is invoked to obtain the free energy of intersection of the two domains. It therefore suffers from the same limitations and approximations, except, of course, that it is not invalid for dilute solutions. It also suffers from limitations arising from the approximations introduced in this section. We should therefore expect no more than qualitative or semi-quantitative agreement with experiment. Some of these approximations have been mitigated in later treatments of the problem, but only by the introduction of greater mathematical complexity.

3.18 The comparison of the theory of dilute solutions with experiment

Osmotic-pressure and light-scattering measurements are both in general agreement with the prediction that a solvent or temperature can be found in which the second virial coefficient is zero.

One important test consists in plotting the second virial coefficient against the reciprocal of the temperature. The relations described above suggest that such plots should be linear in the vicinity of the theta temperature, and that the intercept at $B = 0$ should be independent of the molecular weight and equal to $1/\Theta$. This prediction has been verified for several polymer–solvent systems.

It should be noted that reliable estimates of the second virial coefficient, B, from osmotic-pressure or light-scattering data are only possible if the third virial coefficient, C, is not too large. For this reason effort has been expended in extending the theory to obtain expressions for C. These have resulted in the relation

$$M_2 C = g(M_2 B)^2 \tag{3.49}$$

where g is a parameter that is zero when $B = 0$, but which increases to a limit of 0.25 as B increases. It thus follows that the curvature of plots of Π/ρ_2 versus ρ_2, which depends on C, increases as the initial slope, which depends on B, increases and as we increase the goodness of the solvent. We thus see that estimates of B are more reliable in poor solvents in the vicinity of the theta temperature. For good solvents g may be put equal to its limiting value of 0.25, and it then follows from equations (3.26) and (3.49) that

$$\Pi/\rho_2 = \frac{RT}{M_2}\left(1 + \frac{1}{2}BM_2\rho_2\right)^2 \tag{3.50}$$

and a plot of $\sqrt{(\Pi/\rho_2)}$ versus ρ_2 should then be linear with a slope of $\frac{1}{2}B\sqrt{(RTM_2)}$. It should be emphasised that equations (3.49) and (3.50) apply only to flexible polymers and not to rigid macromolecules such as globular proteins or DNA. It is commonly observed that flexible polymers conform to these relations.

3.19   The conformational entropy of flexible polymers

Several other aspects of the entropy of flexible polymers are of interest, and information concerning them will be required in later chapters.

The first concerns the conformational entropy of an isolated polymer molecule in solution. We have seen that a flexible polymer molecule can adopt any one of a very large number of possible conformations, Ω. $k \ln \Omega$ is the conformational entropy associated with each molecule. It may be viewed as the entropy change per molecule when the molecules, originally in fixed and definite conformations with $\Omega = 1$ (such as is the case if the polymer is in a rigid helical conformation), are allowed to 'melt' so that they are free to adopt any of their possible conformations. An estimate of Ω may be obtained by once more invoking the lattice model and determining in how many ways a polymer molecule may be placed therein irrespective of where the first segment is placed. This is clearly $z(z-1)^{\sigma-2}$ or, to a reasonable approximation, z^σ. Thus the conformational entropy is

$$S_c = \sigma k \ln z \qquad (3.51)$$

The appropriate value of z depends on the flexibility of the polymer, and σ may be taken to be proportional to the molecular weight. For stiff polymers showing little flexibility, z may be close to unity, giving values of S_c close to zero.

The second problem concerns a polymer chain constrained in some way so that only some of its possible conformations are available to it. Such constraints must decrease the conformational entropy. Consider a polymer molecule such that its two ends are fixed in space. It is convenient to suppose that one end is fixed at the origin of a coordinate system and that the other is located in a small volume element, $dx dy dz$, with coordinates x, y, z.

We shall suppose that the proportion of all possible conformations that are available to such a constrained polymer molecule is proportional to the volume of the element, $dx dy dz$, and equal to $W(x, y, x) dx dy dz$, where $W(x, y, z)$ is some distribution function whose value depends on the coordinates, x, y, z. In chapter 4 we shall see that the distribution of end-to-end distances of high-molecular-weight polymers is given to a good approximation by a Gaussian distribution, so that

$$W(x, y, z) = \left(\frac{3}{2\pi\sigma\beta^2}\right)^{3/2} e^{[-3(x^2+y^2+z^2)/2\sigma\beta^2]} \qquad (3.52)$$

This approximate expression is valid as long as the distance between the two ends of the chain, $(x^2 + y^2 + z^2)^{1/2}$, is not too large. In the expression, β is a constant depending on the nature and flexibility of the polymer, and σ is the number of monomer units.

We are now in a position to calculate the entropy of cyclisation; that is the entropy change when one mole of polymer molecules, originally unrestricted, are caused to have their two ends joined to form a circular chain. The number of conformations available before cyclisation is Ω. The number after is obtained from equation (3.52) by putting $x = y = z = 0$ and assuming that both ends are constrained to lie in the same small volume δv; this number is seen to be $\Omega[3/(2\sigma\pi\beta^2)]^{3/2}\delta v$ so that the proportion of the total number of conformations that are available to the cyclic molecule is proportional to $\sigma^{-3/2}$ and the entropy change of cyclisation per mole of polymer is $R \ln \{\Omega[3/(2\sigma\pi\beta^2)]^{3/2}\delta v\} - R \ln \Omega$. This evaluates to $-\frac{3}{2}R \ln \sigma + \frac{3}{2}R \ln [(3/2\pi\beta^2)\delta v^{2/3}]$.

The validity of these expressions has been demonstrated by studies of the equilibrium concentration of cyclic ring-like polymers in reacting systems. It was further shown that the relations are not valid for short chains where the validity of equation (3.52) breaks down.

A third problem concerns the entropy change that occurs when the free end of the chain, originally in a volume element $dxdydz$ at x, y, z, is moved to the point $\alpha_x x$, $\alpha_y y$, $\alpha_z z$ and constrained in a volume element $\alpha_x \alpha_y \alpha_z dxdydz$. This entropy change is clearly given by

$$+ R \ln \left\{ \exp \left[\frac{-3(\alpha_x^2 x^2 + \alpha_y^2 y^2 + \alpha_z^2 z^2)}{2\sigma\beta^2} \right] \alpha_x \alpha_y \alpha_z dxdydz \right\}$$

$$- R \ln \left\{ \exp \left[\frac{-3(x^2 + y^2 + z^2)}{2\sigma\beta^2} \right] dxdydz \right\}$$

per mole of polymer chains. This expression is easily reduced to give an expression for the entropy change of deformation $\Delta S(\alpha_x, \alpha_y, \alpha_z)$ as

$$\Delta S(\alpha_x, \alpha_y, \alpha_z) = -R \left(\frac{3}{2\sigma\beta^2} \right) [(\alpha_x^2 - 1) x^2 + (\alpha_y^2 - 1) y^2 + (\alpha_z^2 - 1) z^2]$$

$$+ R \ln \alpha_x \alpha_y \alpha_z \tag{3.53}$$

If the coefficients, α, are all greater than unity, which corresponds to a 'stretching' of the polymer chains, the entropy change is negative. We shall see in a later chapter that it is this entropy change which lies at the source of the retractive force which is observed when rubber is stretched.

It must be emphasised that equations (3.52) and (3.53) are only valid for chains of high molecular weight and are not valid for oligomers. Furthermore, they are only strictly valid at the theta point, though more elaborate expressions may be derived which are valid for good solvents, or temperatures above the theta temperature.

3.20   Concentration fluctuations

So far in this chapter we have been concerned with the bulk properties of macroscopic portions of solutions of large molecules at equilibrium. These properties are therefore invariant with time. When the system of interest becomes very small (of molecular dimensions) it is observed that the measurable properties exhibit random fluctuations. A well-known example is the Brownian motion of chloroplasts, chlorophyll-containing bodies found in plant tissues, which may be observed under the microscope. A consequence of this random Brownian motion is that the number (or concentration) of particles (or macromolecules) in a small volume element fluctuates with time. The thermodynamic conclusions that we have reached in this chapter remain valid, but only in the sense that they apply to time-average quantities rather than to time-invariant quantities.

An example of interest, which is also important in the theory of light scattering described in a later chapter, is the fluctuations in the concentration of a macromolecule in a small volume, v, of the solution. If ρ_2 is the bulk mass concentration of the solute, the concentration at any instant of time in the small volume will be $\rho_2 + \delta\rho$, where $\delta\rho$ is the instantaneous fluctuation. Since the time-average concentration in this volume is ρ_2 it is clear that the time-average fluctuation, $\langle\delta\rho\rangle$, is zero. However, the time average of the square of the fluctuation, $\langle\delta\rho^2\rangle$, may be finite. Associated with the concentration fluctuations, there are fluctuations also in the free energy, δG, about the mean value, $\langle G\rangle$, and by a similar argument $\langle\delta G\rangle = 0$.

Now the free energy of the small volume of solution is a function of the concentration of the solute, so that there is a relation between the concentration fluctuation, $\delta\rho$, and the free energy fluctuation, δG. This relation may be expressed by expanding the functional relation between G and ρ_2 by a Taylor series about their mean value to obtain

$$\delta G = \left(\frac{\partial G}{\partial \rho_2}\right)\delta\rho + \frac{1}{2}(\delta\rho)^2 \left(\frac{\partial^2 G}{\partial \rho_2^2}\right) + \cdots \tag{3.54}$$

At equilibrium, the free energy is at a minimum, which implies that $\partial G/\partial \rho_2 = 0$. We now suppose that the probability of a concentration fluctuation is given by the Boltzmann distribution to be proportional to $e^{-\delta G/kT}$. This enables us to write an expression for the mean square concentration fluctuation, $\langle\delta\rho^2\rangle$, in the form

$$\langle\delta\rho^2\rangle = \frac{\displaystyle\int_0^\infty (\delta\rho)^2 \exp\left[-\left(\frac{\partial^2 G}{\partial \rho_2^2}\right)(\delta\rho)^2/2kT\right] d(\delta\rho)}{\displaystyle\int_0^\infty \exp\left[-\left(\frac{\partial^2 G}{\partial \rho_2^2}\right)(\delta\rho)^2/2kT\right] d(\delta\rho)} = \frac{kT}{\left(\dfrac{\partial^2 G}{\partial \rho_2^2}\right)}$$

$$\tag{3.55}$$

The differential $\partial^2 G/\partial \rho_2^2$ may be evaluated by differentiating the expression, derived from equation (3.4), $G = n_1 \mu_1 + n_2 \mu_2$, where n_1 and n_2 are the numbers of moles of solvent and solute in the small volume element. The result may be simplified by noting that $v = n_1 \overline{V}_1 + n_2 \overline{V}_2$ which gives a relation between n_1 and n_2 and by noting that the Gibbs-Duhem relation, equation (3.1), implies a relation between μ_1 and μ_2. With these in mind we finally obtain

$$\langle \delta \rho^2 \rangle = - \frac{kT\rho_2 \overline{V}_1}{v \dfrac{\partial \mu_1}{\partial \rho_2}} = \frac{\rho_2}{N_A v \left[\dfrac{1}{M_2} + 2B\rho_2 + \cdots \right]} \tag{3.56}$$

The last part of this expression was obtained by differentiating equation (3.24) with respect to ρ_2, the mass concentration of solute, and noting that for dilute solutions $\overline{V}_1 \simeq V_1^*$.

In equation (3.56), the molar mass M_2 is that of the solute. We thus see that the molecular weight of a macromolecule could be determined from measurements of the concentration fluctuations in a small volume v. If we consider a dilute solution, and retain only the first term of the virial expansion, we see that

$$M_2 = N_A v \langle \delta \rho^2 \rangle / \rho_2 \tag{3.57}$$

In this expression, the mean square concentration fluctuation has been defined to be the time average of the square of the fluctuation observed in a fixed volume of the solution. However, it could as well be defined as the average of the square of the fluctuations observed in a large number of different parts of the same solution and each with a volume v. The two definitions are equivalent in that time averages are equal to ensemble averages.

Various methods may be employed to monitor the concentration fluctuations and to obtain their averages. In principle any property of the solution that varies with the concentration may be used, provided it may be measured with sufficient precision. The measurement of the fluorescent intensity of a fluorescent macromolecule is one method that has been successfully employed.

Problems

3.1 Osmotic-pressure determinations were made on a series of solutions of a globular protein believed to be spherical in shape. The following results were obtained:

Protein concentration (mg/cm³)	Osmotic pressure (Pa)
1	16.00
2	32.10

3	48.35
4	64.70
5	81.15

The temperature at which the measurements were made was $27°C$. Calculate the molecular weight of the protein and its second virial coefficient. Also calculate the excluded volume and compare your result with the partial molar volume assuming that the partial specific volume is $0.75 \text{ cm}^3/\text{g}$. Comment on the results of this comparison.

3.2* Using equation (3.20), derive an expression for the relative chemical potential of the solute using the Gibbs–Duhem relation for a two-component system.

3.3* Derive equation (3.28) from equation (3.27).

3.4* Derive equation (3.32) from equation (3.31).

3.5 A series of solutions of polystyrene of varying molecular weight and dissolved in cyclohexane were cooled until they just became opalescent. The temperatures at which this occurred were recorded, and are given below together with the molecular weight. Assume that the volume of one monomer unit is the same as that of a cyclohexane molecule and estimate the theta temperature of polystyrene in cyclohexane.

$T_c(°C)$	M
31.3	920 000
27.8	182 000
23.9	63 900
19.5	31 500

3.6 Use the answer from problem 3.5 to calculate the second virial coefficient at $40°C$ of each fraction of polystyrene, assuming that the solutions are concentrated and that the density of pure cyclohexane is 0.9.

3.7 The fluctuations in the concentration of a solution of T2 phage DNA were measured over a large number of small volume elements of the solution. The mean square value of $\delta\rho/\rho_2$ was found to be 6.6×10^{-5} when the average concentration of the solution, ρ_2, was $1.9 \times 10^{-7} \text{ g/cm}^3$ and the volume of the elements was $1.5 \times 10^{-5} \text{ cm}^3$. Calculate the molecular weight of the DNA.

4 The configurational statistics of linear polymers

In the theory of dilute solutions of linear polymers discussed in the last chapter we indicated that the chain inhabited a domain. In this chapter we shall be concerned with parameters that describe the size of this domain and the distribution of segments within it.

We also saw in chapter 2 that linear polymers were generally flexible objects on account of the possibility of rotation about the bonds which connect the backbone atoms. For this reason a polymer chain may adopt a very large number of possible conformations. This means that parameters describing the size of the domain must be averages of properties describing the individual conformations. We may view such an average property as the time-average property of an individual polymer chain as measured over a long period of time, or, alternatively and equivalently, as the average at an instant of time of a large number of different chains: the ensemble average.

These average properties are related to the geometrical and thermodynamic properties of the chains: the bond lengths, bond angles and rotational potential-energy characteristics of the chains. They are also related to the nature of the interactions between one monomer unit and another in the chain. We saw in chapter 3 that, in an ideal solvent or at the theta temperature, two polymer domains could freely interpenetrate each other: one domain was not 'aware of the existence' of the other. We shall see in chapter 6 that also at the theta temperature, each part of a single chain is similarly 'unaware of the presence' of the other parts. Polymer chains in circumstances in which this is true are said to be unperturbed, and the elucidation of the conformational statistics is straightforward. For this reason we shall be concerned mainly with the statistics of unperturbed chains.

Under certain circumstances the rotation about the covalent bonds in the polymer backbone may become severely restricted and the chain may adopt a unique (or nearly unique) conformation. Such behavior is exhibited by the helical polynucleotides and by the globular proteins.

In this chapter we shall, for the sake of simplicity, be mainly concerned with the conformation statistics of a particularly simple polymer, polymethylene. The monomer unit of this is a simple methylene group, $-CH_2-$,

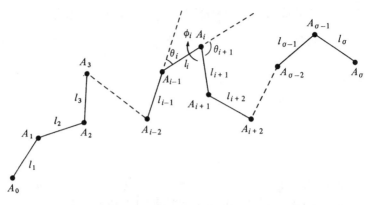

Figure 4.1. Diagram illustrating the definition and numbering of backbone atoms (A_i), bonds and bond length (l_i), bond angles (θ_i) and dihedral angles (ϕ_i). The atoms are numbered starting with 0 at the first in the chain.

so that all the bonds in the backbone are equivalent and all the monomer units (except the two at the ends) are equivalent. The results that we shall discuss can be extended with greater or lesser difficulty to polymers with more complicated structures.

4.1   Bond lengths, bond angles and dihedral angles

In this chapter we shall use the symbol, σ, to denote the number of bonds which join the backbone atoms. We should bear in mind that the number of backbone atoms is then $\sigma + 1$.

It is convenient to label these bonds, starting from one end, by the numbers: $1, 2, 3, \ldots \sigma$. The angle between bond $i - 1$ and bond i we shall call θ_i (there are thus $\sigma - 1$ bond angles to consider). To be more precise we define θ_i to be the supplement of the bond angle as ordinarily defined so that θ_i is zero when bond $i - 1$ and bond i are colinear with atom $i - 1$ between atoms $i - 2$ and i. We define l_i to be the length of bond i.

We also define the *dihedral angle*, ϕ_i, to be the angle between the planes containing bonds $i - 1$ and i on the one hand and bonds i and $i + 1$ on the other. ϕ_i is defined to be zero when all three bonds, $i - 1$, i and $i + 1$, are coplanar. These definitions of l_i, θ_i and ϕ_i are illustrated in figure 4.1.

The bond lengths and bond angles are not perfectly constant, but show small variations, on account of the vibrational motion of the polymer chain. This vibrational motion is, however, generally rapid, and we shall therefore ignore such motion and take the values of the bond lengths and angles to be constant at their mean values.

In polymethylene, all the bonds are of equal length and all the bond angles are the same, but this is not necessarily the case in a more complicated polymer. For these it is sometimes convenient to define an average bond length, l_{av}, as the average of the lengths of all the bonds in the backbone that are associated with one monomer unit.

If the two ends were to be pulled out to their maximum extent (and assuming that the bond angles were all reduced to zero), the distance between the two ends of the chain would be equal to its *contour length*, L. It is clear that $L = \Sigma_i \, l_i$, and, for polymethylene, $L = \sigma l$.

It will sometimes be convenient to represent each bond, i, by a vector, l_i. The specification of l_i then specifies both the length and direction of bond i. Similarly we may represent the vector joining the two ends of the chain in a particular conformation by a vector r. It is then clear that

$$r = \sum_i l_i \tag{4.1}$$

4.2 Conformational parameters and their averages

The simplest average property of interest to us is the mean square end-to-end distance of the chain; that is the average value of the length of the vector r as defined in the previous section. We shall write this average as $\langle r^2 \rangle_0$. The angle brackets indicate time (or ensemble) averages over all possible conformations. The subscript 0 indicates that the average refers to an unperturbed chain under theta conditions.

The mean square end-to-end distance is not directly measurable, and the simplest parameter that can be measured (using light-scattering techniques discussed in chapter 7) is the mean square radius of gyration, $\langle s^2 \rangle_0$. The radius of gyration, s, of a particular conformation may be defined by

$$s^2 = \sum_i w_i s_i^2 \Big/ \sum_i w_i \tag{4.2}$$

where s_i is the distance of the ith monomer unit from the instantaneous centre of mass of the polymer chain, and w_i is the weighting factor associated with the ith unit. As we shall see in chapter 7, for light-scattering experiments, w_i should be taken as the total polarisability, α, of the monomer unit, whereas for X-ray scattering it should be taken as the sum of the atomic numbers of each atom associated with a monomer unit. If all the monomer units are identical, the factors, w_i, cancel out, and we have simply $\langle s^2 \rangle_0 = \langle \Sigma_i \, s_i^2 \rangle / \sigma$. No completely general relation exists between $\langle r^2 \rangle_0$ and $\langle s^2 \rangle_0$, but for high-molecular-weight polymers, for which σ is large, it may be shown that

$$\langle s^2 \rangle_0 = \langle r^2 \rangle_0 / 6 \tag{4.3}$$

This simple relation enables values of $\langle r^2 \rangle_0$ to be calculated from experimental values of $\langle s^2 \rangle_0$. We shall therefore devote our attention to the calculation of $\langle r^2 \rangle_0$.

A further parameter of interest is the *characteristic ratio*. The characteristic ratio of a polymer of σ bonds is written, C_σ, and that of a high-molecular-weight polymer as C_∞. C_σ is defined by

$$C_\sigma = \langle r^2 \rangle_0 / \sigma l_{av}^2 \qquad (4.4)$$

and its significance will emerge below. At this point we shall note that the characteristic ratio of polymethylene dissolved in dodecanol-1 at $140°C$ is 6.7 ± 0.1 (C_∞). It is this experimental value that we shall endeavour to calculate in terms of the bond lengths, bond angles and rotational potential of the polymer chain.

Let us now note that the square of the length of a vector is given by its scalar product with itself so that $r^2 = r \cdot r$. From equation (4.1) we then see that

$$r^2 = \sum_i \sum_j l_i \cdot l_j \qquad (4.5)$$

The double sum in this expression contains σ^2 terms, each of the form $l_i \cdot l_j = l_i l_j \cos \psi_{ij}$, where ψ_{ij} is the angle between bond i and bond j in an instantaneous conformation. If we average over all conformations we see that $\langle r^2 \rangle_0 = \sum_i \sum_j l_i l_j \langle \cos \psi_{ij} \rangle$. Henceforth, for simplicity, we shall assume that all the bonds are of the same length (as is true for polymethylene) so that

$$\langle r^2 \rangle_0 = l^2 \sum_i \sum_j \langle \cos \psi_{ij} \rangle \qquad (4.6)$$

It is illuminating to write out the terms in this sum in the form of a square array as shown in table 4.1. Here we see that each term on the leading diagonal, of which there are σ terms, is equal to $l_i \cdot l_i$. Since l_i is parallel to itself so that $\cos \psi_{ii} = 1$, these terms are all equal to l^2. We may also observe that the terms that lie above the diagonal are all repeated below it. A little thought will then show that we may write

$$\langle r^2 \rangle_0 = \sigma l^2 + 2 l^2 \sum_{1 \leqslant i < j \leqslant \sigma} \langle \cos \psi_{ij} \rangle \qquad (4.7)$$

We may finally observe that $\cos \psi_{i,i+1} = \cos \theta_{i+1}$. The other terms, $\langle \cos \psi_{ij} \rangle$, are trigonometric functions of the various angles θ and ϕ which define the orientation of bond i with respect to bond j.

$l \cos \psi_{ij}$ has an interesting significance: it is equal to the length of the projection of bond i onto the direction of bond j, so that $l \langle \cos \psi_{ij} \rangle$ is equal to the average of this projection. As such, it is a measure of the *correlation* of the direction of bond i with that of bond j. We shall argue below that, if the bonds are sufficiently far apart in the polymer chain, $\langle \cos \psi_{ij} \rangle = 0$, so that

Table 4.1. *Terms in the expansion of* $\overset{\sigma}{\underset{1}{\Sigma}}_i \, \overset{\sigma}{\underset{1}{\Sigma}}_j \cos \psi_{ij}$

	j					
i	1	2	3	4	5	$\ldots \sigma$
1	$\underline{\cos} \, \psi_{11}$	$\cos \psi_{12}$	$\cos \psi_{13}$	$\cos \psi_{14}$	$\cos \psi_{15} \ldots \cos \psi_{1\sigma}$	
2	$\cos \psi_{21}$	$\underline{\cos} \, \psi_{22}$	$\cos \psi_{23}$	$\cos \psi_{24}$	$\cos \psi_{25} \ldots \cos \psi_{2\sigma}$	
3	$\cos \psi_{31}$	$\cos \psi_{32}$	$\underline{\cos} \, \psi_{33}$	$\cos \psi_{34}$	$\cos \psi_{35} \ldots \cos \psi_{3\sigma}$	
4	$\cos \psi_{41}$	$\cos \psi_{42}$	$\cos \psi_{43}$	$\underline{\cos} \, \psi_{44}$	$\cos \psi_{45} \ldots \cos \psi_{4\sigma}$	
5	$\cos \psi_{51}$	$\cos \psi_{52}$	$\cos \psi_{53}$	$\cos \psi_{54}$	$\underline{\cos} \, \psi_{55} \ldots \cos \psi_{5\sigma}$	
$\vdots$						
σ	$\cos \psi_{\sigma 1}$	$\cos \psi_{\sigma 2}$	$\cos \psi_{\sigma 3}$	$\cos \psi_{\sigma 4}$	$\cos \psi_{\sigma 5} \ldots \underline{\cos} \, \psi_{\sigma\sigma}$	

Each element in the square array is one of the terms in the sum on the right-hand side of equation 4.6. The terms in a column originate from the summation over j for a given value of i indicated by the row. The factor l^2 has been omitted for brevity. The terms on the leading diagonal of the form $\cos \psi_{ii}$ are all equal to unity. Each term above the diagonal of the form $\cos \psi_{ij}$ is repeated below the diagonal in the form $\cos \psi_{ji}$.
The terms on the leading diagonal have been underlined.

only a fraction of the terms in equation (4.7) is significantly different from zero.

A further parameter of interest is the *persistence length*, a, of the polymer. This is a characteristic of polymer chain of a certain type, rather than a characteristic, like $\langle r^2 \rangle_0$ or $\langle s^2 \rangle_0$ of a chain with a particular number of monomer residues. To define the persistence length we consider a hypothetic polymer (characterised by given bond lengths, bond angles etc.) of infinite degree of polymerisation. If we consider a bond, i, far from either end of such a chain, the persistence length of the polymer is defined as the average projection of all other bonds j ($j > i$) onto the direction of bond i. Thus

$$a = l \sum_{j=i+1}^{\infty} \langle \cos \psi_{ij} \rangle \tag{4.8}$$

It may be seen that the persistence length is then equal to the sum of all the terms in table 4.1 in the corresponding column (or row) below the diagonal. For flexible polymers only a few of these terms are significant, so that the persistence length is finite even if σ tends to infinity. We shall see that the persistence length is a measure of the flexibility of the polymer chain, which increases as the chain under consideration becomes more rigid and less flexible. It may also be shown that there is a relation between the limiting characteristic ratio, C_∞, and the persistence length:

$$C_\infty = (2a/l) - 1 \tag{4.9}$$

So far we have not considered the details of the averaging process implied in the definition of $\langle r^2 \rangle_0$ etc. According to the Boltzmann distribution, each value of r that enters into the average and which corresponds to a particular conformation should be weighted by a factor $e^{-V\{\theta,\phi\}/kT}$, where $V\{\theta, \phi\}$ is the energy of the conformation. $V\{\theta, \phi\}$ in turn depends on all the angles θ and ϕ which define the conformation. We must therefore write

$$\langle \cos \psi_{ij} \rangle = \frac{\displaystyle\int_0^{2\pi} \cdots \int_0^{2\pi} e^{-V\{\theta,\phi\}/kT} \cos \psi_{ij}\, d\{\theta, \phi\}}{\displaystyle\int_0^{2\pi} \cdots \int_0^{2\pi} e^{-V\{\theta,\phi\}/kT} d\{\theta, \phi\}} \tag{4.10}$$

where we have written $d\{\theta, \phi\}$ to indicate that we must integrate over *all* of the angles θ_i and ϕ_i. This expression looks and indeed is formidable, and progress can only be made by finding simplifying assumptions that can be justified. We shall approach the problem by starting with the simplest possible assumptions and then introducing qualifications into this simple model in order to make it more realistic.

4.3   The random-flight polymer

The simplest possible model for a flexible polymer is the *ideal* or *random-flight* polymer. In this we assume that each bond angle θ_i and each dihedral angle ϕ_i can take any value between 0 and 360° with equal probability so that the energy of every possible conformation is the same. This implies that the directions of every pair of bonds in the chain are uncorrelated and that $\langle \cos \psi_{ij} \rangle = 0$ for $i \neq j$. We then see from equation (4.7) that

$$\langle r^2 \rangle_0 = \sigma l^2 \tag{4.11}$$

and that the characteristic ratio is independent of chain length and equal to unity: far short of the experimental value of 6.7 for polymethylene. We similarly see that the persistence length is equal to l, the bond length.

This model is of course unrealistic for, although the bond angles may exhibit some variation on account of the vibrational motion of the macromolecule, they are constrained to be close to their mean value. We have also seen in chapter 2 that rotation about bonds is accompanied by a change in the energy described by the rotational energy potential functions. The model also tacitly assumes that conformations in which two monomer units occupy the same volume are not excluded and that intramolecular interactions are of no significance. We shall discuss and remove all these limitations in refinements to the model below. Here we shall note that, in a certain sense to be defined below, all polymer chains of sufficient degrees of polymerisation behave like random-coil polymers.

4.4 The freely rotating chain

The most glaring defect in the random-flight model is the assumption of continuously variable bond angles. This is the easiest limitation to remove. In practice we expect the bond angle in polymethylene to be close to the supplement of the tetrahedral angle of about $109°28'$.

If we continue to assume that the energy of the chain is independent of the dihedral angles, so that all ϕ_i can take any value between 0 and $360°$ with equal probability, it is possible to show (see Tanford, 1961, in the further reading section for chapter 4) that $\langle \cos \psi_{i,i+k} \rangle = \cos^k \theta$ where θ is the constant bond angle. We may then see that the sum in equation (4.7) becomes a double geometrical series which may be evaluated exactly to give

$$C_\sigma = \frac{1 + \cos \theta}{1 - \cos \theta} - \frac{2 \cos \theta}{\sigma} (1 - \cos^\sigma \theta)(1 - \cos \theta)^{-2} \qquad (4.12)$$

Since $\cos \theta$ is less than unity, this tends to a limit when σ is large given by

$$C_\infty = \frac{1 + \cos \theta}{1 - \cos \theta} \qquad (4.13)$$

Similarly the persistence length may be shown to be $l/(1 - \cos \theta)$. If θ is small, so that $\cos \theta$ is close to unity, the persistence length becomes very large, and the chain is more rigid than flexible.

For polymethylene, $\cos \theta$ is close to $\frac{1}{3}$, giving a value of C_∞ of 2. The value of C_σ is within 1% of this limiting value for $\sigma > 100$.

4.5 Chains with independent rotational potentials

We have seen in chapter 2 that for n-butane rotation about the central carbon bond caused the energy to vary and to pass through three minima, corresponding to t, g^+ and g^- states, and three maxima, corresponding to eclipsed configurations. It is to be expected that the energy of polymethylene would vary in a similar manner if rotation took place about any of the backbone bonds.

In this section we shall assume that the total rotational energy of the entire chain is the sum of energies associated with rotation about each bond. This implies that the energy associated with bond i depends on ϕ_i, but is independent of any of the other dihedral angles. That is to say, the rotational energy potentials are independent.

We may therefore define an average value of $\cos \phi_i$, $\langle \cos \phi \rangle$, which will be the same for each bond and given by

$$\langle \cos \phi \rangle = \frac{\displaystyle\int_0^{2\pi} e^{-V(\phi)/kT} \cos \phi \, d\phi}{\displaystyle\int_0^{2\pi} e^{-V(\phi)/kT} \, d\phi} \tag{4.14}$$

To proceed further we must relate $\langle \cos \phi \rangle$ to the $\langle \cos \psi_{ij} \rangle$. This can be done via equation (4.10), but the general result is of some complexity (see Tanford, 1961, in further reading section for chapter 4) and will not be quoted here. Having got the result, we may then employ equation (4.7) to obtain $\langle r^2 \rangle_0$. If we make one further assumption, that the rotational potential is symmetric, that is to say $V(\phi) = V(-\phi)$, an explicit expression may be obtained for $\langle r^2 \rangle_0$ (this latter assumption should be valid for polymethylene). The general result is again of some complexity, and we will only quote the limiting value of the characteristic ratio for high degree of polymerisation:

$$C_\infty = \frac{1 + \cos \theta}{1 - \cos \theta} \frac{1 + \langle \cos \phi \rangle}{1 - \langle \cos \phi \rangle} \tag{4.15}$$

The evaluation of C_∞ clearly requires that of $\langle \cos \phi \rangle$. To evaluate this, knowledge of the rotational potential function is needed. This function is rarely known with exactitude. The function for rotation about the carbon-carbon bond in a molecule like ethane has three-fold symmetry and may be approximated by $V(\phi) = \frac{1}{2} V_0 (1 - \cos 3\phi)$ as illustrated in figure 2.2. If this expression is substituted into equation (4.14) it is easily shown that $\langle \cos \phi \rangle = 0$. This means that, if the independent rotational potential is of this form, it would not affect the characteristic ratio, whatever the height of the energy barrier, V_0. A more realistic assumption is that the potential resembles that of n-butane as illustrated in figure 2.5, in which there are again three minima but with different energies. If we assume that the potential function can be represented by the expression given in the caption to figure 2.5 we find that $\langle \cos \phi \rangle = 0.213$. This in turn results in a value of C_∞ of 3.18. It may be pointed out that $\langle \cos \phi \rangle$ is insensitive to the value assumed for ΔV_{tg}, the energy difference between the t and g states. In order to raise the characteristic ratio of polymethylene to the experimental value of 6.7, a quite unrealistically high value of ΔV_{tg} would have to be assumed. It is thus clear that the assumption of independent rotational potentials is inadequate to account for the experimental values of the characteristic ratio.

4.6   The rotational isomeric model

In figure 4.2 we have plotted $e^{-V(\phi)/RT} \cos \phi$ against ϕ using the expression for $V(\phi)$ illustrated in figure 2.5. This figure shows the relative

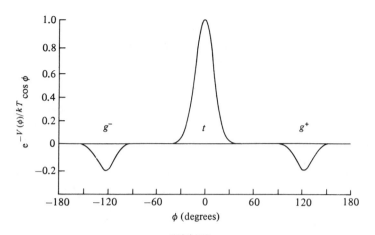

Figure 4.2. Plot of $e^{-V(\phi)/kT} \cos \phi$ versus ϕ. The form of $V(\phi)$ assumed was the same as that used to construct figure 2.5. The diagram illustrates the relative weighting of different values of ϕ to the average $\langle \cos \phi \rangle$.

contribution to $\langle \cos \phi \rangle$ from various values of ϕ (see equation (4.14)). It may be seen that most of the contribution arises from values of ϕ close to the minima in the potential function.

This suggests that a good approximation to the integrals in equation (4.14) could be achieved by replacing them by the sum of three terms, each corresponding to one of the three minima. To do this we define a statistical weight, w_i, for each minima by

$$w_i = e^{-V(\phi_i)/RT} \qquad (4.16)$$

where ϕ_i is the value of the angle ϕ at which the minimum in the potential function occurs. We then write

$$\langle \cos \phi \rangle = \frac{w_t \cos \phi_t + w_{g^+} \cos \phi_{g^+} + w_{g^-} \cos \phi_{g^-}}{w_t + w_{g^+} + w_{g^-}} \qquad (4.17)$$

In n-butane, the t state occurs at $\phi = 0$ and the two g states at $\phi = \pm 120°$. The energy difference $V(\phi_g) - V(\phi_t) = \Delta V_{tg}$ has been estimated to be 2.1 kJ/mol. Assuming these values hold also for polymethylene, we obtain from equation (4.17) an estimate for $\langle \cos \phi \rangle$ of 0.219 and for C_∞ of 3.12. These values for $\langle \cos \phi \rangle$ and C_∞ are not significantly different from those obtained in the previous section using the full integral for $\langle \cos \phi \rangle$ given by equation (4.14).

This model, in which only rotational states corresponding to the minima of the potential function are considered, is termed the *rotational isomeric model*. We see that estimates of $\langle \cos \phi \rangle$ obtained with its use are barely distinguishable from the more exact treatment.

The rotational isomeric model implies that the conformation of the entire chain may be specified by indicating which rotational state, t, g^+ or g^-, each bond has adopted. Thus the state of the chain could be specified by a sequence of symbols such as $ttg^+tg^-g^-t \ldots$ Of course we do not suppose that the dihedral angles take on values exactly equal to the angle corresponding to an energy minimum. Rather we suppose that they take on values close to this minimum angle and distributed equally on either side of it. We must also bear in mind that there is associated with this distribution of angles about the minimum a certain entropy, which should in principle be allowed for in the definition of the weighting factors, w_i. If, however, the 'wells' in the potential energy diagram are of equal width, as is approximately the case in reality, these entropy factors will cancel. We are therefore justified in ignoring this complication.

It is true that the introduction of the rotational isomeric model has not resulted in a significant improvement of our estimate of the characteristic ratio of polymethylene, but, nevertheless, it will open the way to the next level of approximation, in which the condition that the rotational potentials are independent is relaxed. It does this by permitting a vast simplification of the mathematics by allowing the integrals such as are involved in equation (4.14) to be replaced by sums of a small number of terms.

4.7   Interdependent rotational potentials

We have seen in section 2.12 that, in n-pentane, adjacent g^+ and g^- states are statistically improbable on account of the operation of exclusion forces acting between the two terminal methyl groups. This means that the statistical weight w associated with the rotational isomeric state of one bond is dependent on the rotational isomeric state of its neighbours. Not only may the energies at the minima of the energy diagram depend on the state of the neighbouring bonds, but also the actual values of the dihedral angles at which these minima occur may be affected. Such dependencies may be depicted in conformational maps as described in section 2.12 and illustrated in figure 2.6.

We summarise these observations by saying that each bond in the polymer backbone interacts with its neighbours. We shall suppose that this interaction is limited to nearest neighbours, and that the state of bond i is affected by the state of bond $i - 1$. Similarly the state of bond $i + 1$ is affected by the state of bond i etc. It is true that the state of bond i is also affected by the state of its other nearest neighbour, bond $i + 1$, but no loss of generality results if we attribute the results of this interaction to the effect of the state of bond i on that of bond $i + 1$. Thus the state of each bond depends on the state of its predecessor along the chain.

This assumption of nearest-neighbour interactions severely complicates the mathematics involved in deriving an expression for the characteristic ratio, but does not render this programme unfeasible. A complete solution to the problem is available, but it requires an understanding of matrix algebra for its understanding, and even then the expression for C_∞ is of some complexity. We shall therefore only consider the results of the application of the theory to polymethylene.

It has not proved possible to define with exactitude either the values of the dihedral angles associated with the rotational states of polymethylene or of the associated energies. Nevertheless, estimates of these parameters, which lie within reasonable narrow ranges, have been obtained from spectroscopic investigations on simple n-alkanes and the semi-empirical calculations mentioned in section 2.12 and which are based on a consideration of energies of interaction between the various groups and atoms involved.

If reasonable estimates of these parameters obtained in this manner are used to calculate the characteristic ratio, C_∞, a value close to the experimental value of 6.7 is obtained.

These selfsame reasonable values of the parameters also lead to an estimate of the temperature coefficient of C_∞, which is also close to the experimental value of $1.0 \times 10^{-3}/K$. We may therefore consider that the considerable effort that has gone into the development of the general theory has been vindicated in the case of polymethylene chains.

4.8 Application to other polymers

Polymethylene is perhaps the simplest polymer: it is a homopolymer; its monomer unit consists of but three atoms, $-CH_2-$; the bonds in the backbone are all equivalent; the various bonds are essentially non-polar; strong interactions between monomer units are absent. Moreover considerable experimental data are available concerning the rotational potentials in oligomeric homologues.

Many polymers of interest fall short of these ideals in one or more ways. In both polypeptides and polynucleotides there are several different sorts of bonds in the polymer backbone about which free rotation may take place. Although this complicates still further the expression for the characteristic ratio, it is not outside the scope of the general theory. Similarly, the loss of symmetry in the rotation potentials present in the stereoregular polymers may be taken into account. More serious are the finite volumes of the side groups in, for instance, polypeptides and polynucleotides and the possibility of strong attractive interactions between them. Ways of coping with this difficulty are outlined below.

A drawback to the application of the theory is the lack of adequate knowledge concerning the energies of the various minima and the corresponding dihedral angles and calculations based on the calculation of energies of interaction leave considerable scope for error. Some information concerning the positions of the energy minima defined in terms of the dihedral angles may be obtained from the analysis by X-ray crystallographic methods of crystals of simple monomers or oligomers, but one must always bear in mind that a molecule in a crystal is subject to constraints arising from its interaction with its neighbours in the lattice that may not be present in solution. These constraints may distort the dihedral angles.

4.9   The excluded-volume effect

We have remarked that in n-pentane the g^+g^- state leads to a close approach of the two terminal methyl groups and that such analogous states in polymethylene must be excluded (or at least but lightly weighted) in the calculation of the mean square end-to-end distance. The incorporation of interdependent rotational potentials at the level of nearest-neighbour interactions is able to cope with this pentane effect. There must, nevertheless, exist conformations of the polymer in which similar close approaches between more distant neighbours would occur and which should be omitted in the averaging process. Such effects are aggravated if the monomer units carry bulky side groups as in polynucleotides and polypeptides. There is at present no general or rigorous method of dealing with this effect which is referred to as the *excluded-volume effect*.

If such conformations involving close approaches of distant monomer units were to be omitted from the averaging process, the result would be to increase the calculated mean square end-to-end distance. This may be seen by the following argument. Consider all conformations in which a particular pair of monomer units are in close proximity. These two residues and all others situated between them along the chain then constitute a closed loop as illustrated in figure 4.3. The mean square end-to-end distance of those conformations of the polymer under consideration is then determined as if it were a shorter chain with those residues in the loop excised. Since the mean square end-to-end distance of a chain increases with σ, the number of residues it contains, we see that $\langle r^2 \rangle$ for these shorter chains is less than that for the whole chain taking into account all conformations. Thus, if the conformations involving loops of the kind described are eliminated from the average, the value of $\langle r^2 \rangle$ using only those conformations that remain must be increased. We thus see that the result of the excluded-volume effect is to increase the mean square end-to-end distance to a value greater than would pertain in its absence.

An opposite effect may occur if strong intramolecular interactions occur between distant residues along the chain. These will favour conformations in which residues are in close proximity, and result in a reduction of $\langle r^2 \rangle$. The probability of such strong interactions depends on the temperature and on the nature of the solvent. We shall see in chapter 6 that arguments may be

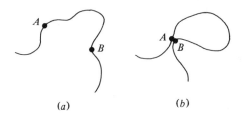

(a)   (b)

Figure 4.3. Illustration of the formation of a closed loop in a polymer chain when two residues are in close proximity. The mean square end-to-end distance of the conformations in which there is a loop (b) is smaller than that of conformations in which the loop is absent (a).

adduced for supposing that they cancel the excluded-volume effect at the theta temperature. For the present it must be strongly emphasised that it is only at the theta temperature (or in a theta solvent) that agreement between experimental values of the mean square end-to-end distance based upon determinations of the mean square radius of gyration and theoretical values calculated by the methods outlined above can be expected. The quest for theta conditions is paramount.

We shall argue in the next section that for any flexible polymer of sufficiently great degree of polymerisation, σ, the unperturbed mean square end-to-end distance, $\langle r^2 \rangle_0$, is proportional to σ and hence to the molecular weight, so that we may write

$$\langle r^2 \rangle_0 = \sigma \beta^2 \qquad (4.18)$$

where β is constant, independent of σ, and depends only on the bond lengths, bond angles and other molecular parameters. We may therefore formally take into account the excluded-volume effect and other long range intramolecular interactions by writing

$$\langle r^2 \rangle = \sigma \beta^2 \alpha^2 \qquad (4.19)$$

where $\langle r^2 \rangle$ is the mean square end-to-end distance of the perturbed polymer chain so that $\alpha^2 = \langle r^2 \rangle / \langle r^2 \rangle_0$, and $\alpha = 1$ under theta conditions. No rigorous method is available for relating α to the molecular properties of the polymer, but in chapter 6 we shall outline an approximate theory which has the merit of being consistent with many experimental results in a semi-quantitative sense. In particular, this theory suggests that α is proportional to $\sigma^{0.1}$ when α is large. This occurs at high temperatures or in very good solvents. Under these circumstances we see that $\langle r^2 \rangle$ is proportional to $\sigma^{1.2}$.

Several attempts have been made to justify this relation between $\langle r^2 \rangle$ and σ in a more rigorous, or at least a different way. One sort of attempt is to employ a high speed digital computer to consider a large number of possible conformations defined by a sequence of rotational states. The computer is then instructed to ascertain which of these conformations must be rejected because of the excluded-volume effect and to compute $\langle r^2 \rangle$ for those remaining. In passing it is interesting to note that a large proportion of the possible conformations must be rejected in this way.

Another approach to the problem is to enumerate systematically all possible conformations, an approach limited to relatively short chains. Happily, the majority of results obtained by both approaches are in reasonable agreement and suggest that $\langle r^2 \rangle$ is proportional to $\sigma^{1.2}$, the relation suggested by the thermodynamic approach outlined in chapter 6. It should be recognised however that the problem has not been solved to the satisfaction of all investigators.

4.10   Kuhn statistical segments

We have seen in sections 4.3 to 4.6 that the limiting characteristic ratios of the model polymers discussed therein are all independent of the degree of polymerisation. We thus see that, as far as these models are concerned, a relation of the form of equation (4.16) is valid and the unperturbed mean square end-to-end distance is proportional to σ. Kuhn was the first to note that this proposition should be true for any flexible polymer provided σ is large enough. He pointed out that the correlation between the direction of bond i and bond $i + k$ must fall to zero as k increases. It is true that in very stiff polymers (such as double-helical DNA) k may need to become very large before $\langle \cos \psi_{i,i+k} \rangle = 0$, but in principle it will do so.

Suppose now that we take the first k monomer units of a long polymer chain (taking k to be sufficiently large that $\langle \cos \psi_{i,i+k} \rangle = 0$) counting from one end of the chain and call this set of units a *statistical segment.* Let us suppose that the mean square end-to-end distance (in the unperturbed state) of this statistical segment is $\langle r'^2 \rangle$; it will depend on k and on the bond lengths etc. in an unspecified manner. We now go on to divide the whole polymer chain up into σ' such segments such that $k\sigma' = \sigma$. By hypothesis, there is no correlation between the directions of the vectors joining the ends of the various statistical segments. This implies that the sequence of statistical segments may be treated as a random-flight polymer in which each 'bond' is replaced by a statistical segment of length $\langle r'^2 \rangle^{1/2}$. We then see that the end-to-end distance of the whole chain of σ' segments, $\langle r^2 \rangle_0$, is equal to $\sigma' \langle r'^2 \rangle$. If we then put $\beta^2 = \langle r'^2 \rangle / k$ we obtain equation (4.18), which is thereby justified. The only condition is that the minimum value of k as defined above would be less than $\sigma/2$.

4.11 Helical polymers

So far we have dealt with polymer chains in which there is at least a modicum of flexibility; that is to say that at least some of the bonds can exist in more than one rotational state with finite probability. Such polymers are termed *random coils*.

An interesting situation prevails if the probabilities of all but one of the rotational states of each residue become vanishingly small on account of short range interactions between near neighbours. From a theoretical point of view this will occur at temperatures approaching the absolute zero for all polymers. This can be seen in the case of polymethylene where the probability of a bond being in a *t* state is proportional to $e^{-V(\phi_t)/kT}$. As T falls to zero, this probability becomes equal to unity. A more realistic case is afforded by certain synthetic polypeptides such as polyphenylalanine. In these a hydrogen bond may be formed between the carbonyl oxygen atom of one amino acid with the amide hydrogen of the amino acid preceding it by four residues along the chain; interactions between the amino-acid side chains may also play their role. In double-helical DNA there are hydrogen bonds between the heterocyclic bases of one polynucleotide chain and those on another; such hydrogen-bonded pairs of bases are stacked on top of each other in such a way that they interact with their neighbours. The general result of such inter-action is that corresponding bonds in the different monomer units are forced into equivalent rotational states.

A geometrical consequence of this uniformity of rotational states is that the polymer becomes a helix; that is to say corresponding atoms in the different monomer units lie on a helix. A helix, it is worth noting, has a screw axis of symmetry. If we take a very long (infinitely long) helix and rotate it through an angle θ_h about its axis and then move it along its axis by a certain distance d_h, it is brought into coincidence with itself. According to the direction that this rotation must be made, the helix may be either left handed or right handed, just as there may be screws with left-hand or right-hand threads. If a polymethylene chain is constrained to adopt a configuration in which each bond is in a *t* state, its end-to-end distance is as long as it can be and equal to $\sigma l \cos \frac{1}{2}\theta$. In this state the chain is in a zig-zag conformation which is a helix with $\theta_h = 180°$.

If a polymer in a helical conformation is heated, or the nature of the solvent is changed in such a way as to decrease the intramolecular interactions, the probability of a bond being in an alternative rotational state is increased. The polymer then reverts to a random-coil conformation and the helix is said to melt. Such helix–coil transitions are discussed more fully in the next chapter.

Such helical polymers resemble rigid rods. In so far as this is the case and the rod is quite rigid and there are no kinks in it, the end-to-end distance is proportional to the number of monomer units or the degree of polymerisation, σ. The radius of gyration, s, is given in terms of the length, L, of the rod and its diameter, d, by the relation

$$s^2 = \frac{d^2}{8} + \frac{L^2}{12} \tag{4.20}$$

If the length is large compared to the radius this reduces to $s^2 = L^2/12$.

4.12   The worm-like coil

We have supposed that a molecule like double-helical DNA is a rigid rod-like object. Unfortunately such a supposition does not give a realistic picture of such helical molecules, and two factors both of which give rise to an element of flexibility must be considered. The first of these factors is only of importance at temperatures such that the helix is beginning to melt. Under these circumstances there will be a finite probability that any given bond in the polymer backbone will be in a rotational state which is not consistent with the regular helix. When this occurs the rod will be kinked at this position so that the two portions of the rod which the bond joins will be free to rotate relative to one another. If there are several such kinks (at random positions along the length of the rod) the molecule will resemble a random coil more than a rigid rod.

Let us suppose however that the temperature is sufficiently low that such kinks have a negligible probability of occurrence. Even under such circumstances, the second factor may be important. This depends upon the fact that the minima in the rotational potential energy functions are not infinitely sharp. That is to say small deviations from the nominal values of the dihedral angles can occur with but small cost in energy. This means that the otherwise rigid rod is to a small extent flexible. The effect of this is not likely to be apparent in short rods, but a long rod will again behave more like a random coil.

To account for such effects in a semi-quantitative manner, the *worm-like coil* model was proposed by Porod and Kratky. Let us first suppose that the rod is divided into a number, n, of short segments, each of length, h, such that $nh = L$, the length of the rod. Any segment, because of the bending effect, may be inclined to its neighbour at an angle θ, but the direction of bending, we will suppose, is random as illustrated in figure 4.4. We suppose that the time average of the cosine of any one of these angles, $\langle \cos \theta \rangle$, is finite and that the same value of $\langle \cos \theta \rangle$ pertains to each pair of adjacent segments.

Defined in this way the chain of segments resembles the freely rotating

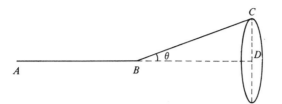

Figure 4.4. *AB* and *BC* are two short segments of a rod that are
joined by a stiff hinge. The inclination of *AB* to *BC* is defined by the
angle θ, but the direction of bending is random so that the end *B*
may lie anywhere on the circle indicated with equal probability. The
length of *BD* represents the projection of *BC* onto the direction of
AB and is equal to *BC* cos θ.

chain discussed in section 4.4. $\langle \cos \theta \rangle$ here plays the role of the cosine of
the bond angle in the freely rotating chain and h the role of the bond length.
    The mean square end-to-end distance of the chain of segments is then given
by equation (4.12) to be

$$\langle r^2 \rangle = nh^2 \left[\frac{1 + \langle \cos \theta \rangle}{1 - \langle \cos \theta \rangle} - \frac{2 \langle \cos \theta \rangle}{n} (1 - \langle \cos \theta \rangle^n)(1 - \langle \cos \theta \rangle)^{-2} \right]$$

(4.21)

We are not able to use the approximation used in section 4.4 to derive
equation (4.13) from (4.12) because in the present situation $\langle \cos \theta \rangle$ is
close to unity. The persistence length of the chain of segments is given by
$h/(1 - \langle \cos \theta \rangle)$. We have had previous occasion to mention that the persistence length of a polymer is a measure of its rigidity, a large persistence
length indicating a relatively stiff and inflexible coil. This interpretation of
persistence length is plainly consistent with the identification of the chain of
segments with a freely rotating coil, for a large value of $\langle \cos \theta \rangle$ implies both a
large persistence length and a small tendency for the rod to bend.
    We now suppose that the length of each segment, h, tends to zero so that,
at the same time, the number of segments, n, tends to infinity, in such a way
that the product, nh, remains equal to L. At the same time, we also suppose
the persistence length, a, remains constant, and $\langle \cos \theta \rangle$ tends to unity as θ
gets smaller and smaller. In this limit of infinitesimal θ, the factor $(1 - \langle \cos \theta \rangle)$
may be replaced by $\frac{1}{2} \langle \theta^2 \rangle$ so that $a = 2h/\langle \theta^2 \rangle$. It is also possible to show that
$\langle \cos \theta \rangle^n$ becomes equal to $e^{-L/a}$. Substituting these limiting values into equation (4.21) then results in the limiting value of the mean square end-to-end
distance of the worm-like coil given by

$$\langle r^2 \rangle / L^2 = 2p(1 - p + p\, e^{-1/p})$$

(4.22)

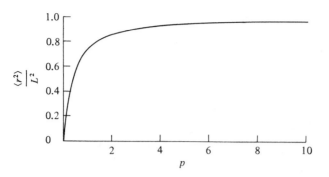

Figure 4.5. A plot of $\langle r^2 \rangle / L^2$ versus p according to equation (4.20). p represents the ratio of the persistence length of a worm-like coil to the contour length, L. $\langle r^2 \rangle$ is the mean square end-to-end distance.

where p is the ratio of the persistence length, a, to the length of the 'rod' from which the worm-like coil was derived, so that $p = a/L$.

For very long rods or very flexible rods, $a \ll L$, $p \ll 1$, and equation (4.22) reduces to $\langle r^2 \rangle = 2aL$. This is the expression appropriate for the mean square end-to-end distance of a freely rotating chain of high degree of polymerisation if L is equated to the contour length and, as is implicit above, a is much larger than the 'bond length', h.

On the other hand, for very short or very inflexible rods, $a \gg L, p \gg 1$, and equation (4.22) reduces to $\langle r^2 \rangle = L^2$. This is the expression expected for a perfectly rigid rod of length, L. The behaviour of $\langle r^2 \rangle / L^2$ for intermediate values of p is exhibited in figure 4.5.

We thus see that the worm-like coil can resemble a random coil at one extreme and a rigid rod at the other, as well as objects of intermediate flexibility, according to the value of the parameter, p, that is used to specify its characteristics. It is also possible to derive an expression for the radius of gyration of the worm-like coil in terms of the parameter, p. It is thus possible to estimate values of the persistence length from measurements of the radius of gyration of macromolecules such as DNA for which the worm-like coil is an adequate model, provided the contour length is known.

Further insight into the significance of the persistence length of the worm-like coil may be gleaned by noting that the introduction of curvature in a stiff rod involves the expenditure of energy. According to Hooke's law, the energy required to bend a segment of it of length $2h$ until the ends of the segment are inclined at an angle to each other is $Z\theta^2/2h$, where Z is the flexural elastic force constant. The Boltzmann distribution then implies that the probability of such a bend occurring is proportional to $e^{-Z\theta^2/2hkT}$. From this relation we may calculate the mean value of θ^2 and hence the persistence length. Thus we may show that the persistence length $a = Z/kT$. From this result it is to be expected that the persistence length decreases at higher temperatures so that the coil becomes more flexible and $\langle r^2 \rangle$ decreases.

4.13 Compact globular polymers

We have seen that in a random coil intramolecular interactions are minimal, but that strong short range interactions between neighbouring residues along the chain can cause the polymer to adopt a unique helical conformation. Another possibility exists for the formation of a unique conformation if strong interactions between more distant residues are possible. Under these circumstances the polymer chain may coil up into a compact coil (though not a random coil) with a defined tertiary structure in which each monomer unit has a fixed position and rotation about the bonds of the backbone is severely inhibited.

The globular proteins are examples *par excellence* of polymers with the capacity to do this, though certain nucleic acids such as transfer RNA can also form a tertiary structure. Both these biopolymers are heteropolymers and both are also capable of adopting helical secondary structures as well. In many globular proteins there are regions of the polypeptide chain that have adopted a helical structure interspersed with other regions with no defined secondary structure. Also in many globular proteins there are covalent cross links between pairs of cysteine residues.

The exact reasons for the stabilisation of globular proteins is incompletely understood. One important generalisation which has emerged from the analysis of the three-dimensional structure of crystals of globular proteins by X-ray diffraction techniques is that there is a tendency for the amino acids with polar side groups to be found at the surface of the globular structure and for those with non-polar side groups to be found in the interior. This has suggested that part of the free energy of stabilisation originates from the 'hydrophobic bond'. It is likely however that other sorts of secondary interaction, particularly hydrogen bonds, also play an important role.

If the temperature of a solution of a globular protein is raised, or the nature of the solvent changed, the tertiary structure is generally disrupted, and the molecules revert to the random-coil state. This process is termed *denaturation*. This process is believed to be reversible so that the random-coil polypeptide can be induced to return to its globular tertiary structure by an appropriate change in the conditions. It is believed that the exact architecture of the unique tertiary structure is determined by the primary structure, that is by the sequence of amino-acid residues along the polymer chain. This implies that it should be possible to predict the tertiary structure from the primary structure, but so far the rules that would permit this to be done remain elusive.

If we consider a polypeptide chain with a defined, though arbitrary, sequence of amino-acid residues, any tertiary structure that it may adopt is subject to the following general constraints:

(*a*) It must, if it is globular, be contained in an approximately spherical domain.

(*b*) There should be little or no free space within the domain.

(*c*) The preponderance of non-polar amino-acid residues should be in the interior of the domain, and the preponderance of polar residues at its surface (though for certain proteins normally existing in a non-polar environment, as in a membrane, this may not be so).

(*d*) The bond angles and bond lengths both in the polymer backbone and in the side groups should not be grossly distorted.

(*e*) The various dihedral angles should, as a generality, be not severely displaced from their positions of minimum energy.

(*f*) Intramolecular interactions including hydrogen-bond formation should be maximised.

These constraints are so severe that it seems likely that only a few conformations, if any, which simultaneously satisfy them all are available to an arbitrary polypeptide.

Some confusion is sometimes engendered by referring to the structure of a globular protein as being a random coil. A coil it may be, but random it certainly is not, and in no way does it resemble the flexible random coils on whose statistical properties we have dwelt in previous sections of this chapter. The conformation is unique (almost) and all (nearly) of the molecules are to be found in it. It would be better to say that the structure was an irregular coil to draw attention to the fact that the polymer chain is arranged in no obvious pattern. It should also be noted that when the tertiary structure reverts to a random coil as the protein is denatured, there is a large increase in conformational entropy.

The domain of a globular protein can sometimes be approximated by a sphere, or, with a little more elaboration, by an ellipsoid of revolution. Ellipsoids of revolution are generated by rotating an ellipse about either its minor axis or its major axis. In the former case an oblate ellipsoid of revolution is generated; this is a flattened disc-like figure. In the latter case a prolate ellipsoid is generated; this is an elongated cigar-like figure. Such figures must be treated as rather gross approximations since the real domain of a globular protein is complicated and irregular possibly with crevices and certainly with a 'knobbly' surface. I have attempted to depict such a domain, together with the ellipsoid that might approximate to it, in figure 4.6.

The unique character of a globular protein implies that it has a well-defined radius of gyration, and this is open to determination from light-scattering experiments. If information is available concerning the axial ratio (which determines the shape) of the ellipse that best approximates the domain, the

Figure 4.6. Illustrative representation in two dimensions of the shape of the domain of a globular protein. The circle represents a sphere that might be taken to represent an approximation to the domain.

experimental value of the radius of gyration can be used to derive an approximation to the size of the domain. The end-to-end distance of a globular protein, although well defined, is not determined by the radius of gyration, and is a parameter of little interest.

It is interesting to compare the radii of gyration of a polymer when it is (*a*) in the random-coil state, (*b*) in a helical rod-like state and (*c*) in a compact spherical state. As an illustrative example we have taken a polymethylene chain of degree of polymerisation 5000 and assumed that the excluded volume of a methylene group is 0.023 nm^3 and that the length of a carbon–carbon bond in the polymer backbone is 0.11 nm. We have calculated the radius of gyration of the unperturbed random coil; that of a hypothetical rod-like form in which each bond in the backbone is in the *trans* configuration; and that of a compact sphere in which all the monomer units are packed into a spherical domain. The results of these calculations are given in table 4.2. They imply that the denaturation of a globular structure is accompanied by an increase in the radius of gyration, whereas the melting of a helical structure is accompanied by a decrease in the radius of gyration. Such predictions are consistent with what is observed when globular proteins are denatured or helical polynucleotides or polypeptides melt.

4.14 Distribution functions for flexible polymer chains

We have seen that the end-to-end distance of a flexible linear polymer chain may take many different values and we have seen how the square of the average of this quantity may be calculated. The question arises as to how the end-to-end distances are distributed. We note in passing that such a

Table 4.2. *Radii of gyration of random-coil rod-like and globular forms of a polymer*

	Radius of gyration (nm)
Random coil	8.2
Rigid rod	130
Compact sphere	2.3

Comparison of the radii of gyration (in nm) of a polymethylene chain of degree of polymerisation $\sigma = 5000$ in the unperturbed random-coil form; in a hypothetical helical form in which all bonds are the *trans* configuration; and in a hypothetical compact globular form. It has been assumed that the carbon–carbon bond length is 0.11 nm, that the excluded volume of the methylene unit is 0.023 nm^3 and that the characteristic ratio of the random coil is 6.7.

question has no significance for helical or globular conformations, for in these the end-to-end distance is unique.

Suppose that one end of a polymer chain is anchored at the origin of a coordinate system. As the chain adopts all possible conformations, we may ask in what proportion of these does the other end lie in a small volume element situated at the point with coordinates x, y, z and of volume $dx\, dy\, dz$? It may be supposed that the proportion is proportional to the volume $dx\, dy\, dz$ and to depend on x, y, z. We may thus define a distribution function, $W(x, y, z)$, such that the ratio of the number of conformations with the end of the chain in the volume element to the total number of conformations is $W(x, y, z)\, dx\, dy\, dz$.

The following argument suggests (though does not prove) that this distribution is Gaussian if the chain is a very long random-flight polymer. First suppose that the proportion of all conformations such that the x coordinate of the free end lies in the interval x to $x + dx$ is $w(x)\, dx$. The crux of this argument is to note that, for a very long chain, this proportion is independent of the distribution in the y and z directions. This is so because there is no preferred direction for the end-to-end vector provided the chain is long enough; if it is not long enough the direction of the first bond will impose a bias on the direction of this vector (think, for instance, of the distribution of a chain of only two links).

It then follows that the proportion of all conformations with the chain end in the interval y to $y + dy$ is given by the same function as $w(y)\, dy$. Similarly for the z direction, the proportion is $w(z)\, dz$. When then see that the proportion of all conformations such that the free end lies simultaneously in all three of these ranges is $w(x)\, w(y)\, w(z)\, dx\, dy\, dz$. Now we have already noted that there is no preferred direction for the end-to-end vector; it therefore follows that this expression depends only on the distance, $(x^2 + y^2 + z^2)^{1/2}$, of the volume element from the origin, and not on the direction. This implies that $w(x) \, w(y) \, w(z) = W(x^2 + y^2 + z^2)$. The only form of the functions $w(x)$ that is consistent with this is $w(x) = b \, e^{-cx^2}$, where b and c are constants, which remain to be determined. We thus see that

$$W(x, y, z) = b^3 \, e^{-c(x^2 + y^2 + z^2)} \tag{4.23}$$

This function defines the distribution of the position of one end of the polymer relative to the other in space. Of equal interest is the distribution of the end-to-end distance, r, irrespective of direction. We thus define the distribution function $\phi(r)$ such that $\phi(r) \, dr$ is the proportion of all conformations such that the two ends are separated by a distance, r. $\phi(r)$ is easily derived from $w(x, y, z)$ as follows.

If one end of the chain is again fixed at the origin, and the other end is at a distance which lies within the range r to $r + dr$, this free end must be within a spherical shell of radius r and thickness dr centred on the origin. The volume of this shell is $4\pi r^2 \, dr$. It then follows that the proportion of all conformations with the free end in the shell is given by $4\pi^2 b^3 \, e^{-cr^2} \, dr$ so that $\phi(r) = 4\pi r^2 b^3 \, e^{-cr^2}$. Clearly the integral of $\phi(r)$ over all values of r from zero to infinity must be unity since all conformations are encompassed by this summation. We therefore see that the constant $b = (c/\pi)^{1/2}$. In order to evaluate the other constant, c, we may evaluate the mean square of r for the distribution and equate the result to $\langle r^2 \rangle_0$. We thus find that $c = 3/2\langle r^2 \rangle_0$ and

$$W(x, y, z) = \left(\frac{3}{2\pi \langle r^2 \rangle_0} \right)^{3/2} e^{-[3(x^2 + y^2 + z^2)/2\langle r^2 \rangle_0]} \tag{4.24}$$

$$\phi(r) = 4\pi r^2 \left(\frac{3}{2\pi \langle r^2 \rangle_0} \right)^{3/2} e^{-(3r^2/2\langle r^2 \rangle_0)} \tag{4.25}$$

The argument we have presented above does not claim to be rigorous. Nevertheless, the results enshrined in equations (4.24) and (4.25) can be shown to apply rigorously in the limit to polymers of infinite degree of polymerisation. Real polymers, for which these equations represent to a reasonable approximation their distributions of end-to-end distances, are sometimes referred to as *Gaussian chains*. For chains of finite length these equations must be invalid for values of r greater than the contour length because they imply a finite proportion of chains with an end-to-end separation greater than this.

The derivation of exact expressions for chains of finite length requires a detailed statistical treatment, and the results depend on the details of the restrictions on chain flexibility that apply. The problem has been solved exactly for random-flight polymers of finite length, but both the argument and the resulting expression are complicated. An approximation to the exact solution that is sometimes quoted is the so called Langevin distribution:

$$\phi(r) = c \exp \left[-\frac{1}{L} \int_0^r \mathcal{L}^{-1}\left(\frac{r}{\sigma L}\right) \mathrm{d}r \right] r^2 \tag{4.26}$$

In this expression, c is a constant whose value must be chosen to make the integral of $\phi(r) \, \mathrm{d}r$ equal to unity. $\mathcal{L}^{-1}(x)$ is the inverse Langevin function of x, that is to say it is equal to the value of y such that $\coth y - (1/y) = x$. L is the contour length.

If the exact solution, the Langevin distribution and the Gaussian distribution are compared for chains of different length, obvious differences may be perceived between the three distributions for very short chains; but, for chains with $\sigma > 20$, the three distributions are closely similar except for very large values of r approaching the contour length.

Approximate distribution functions have also been derived for chains incorporating various restrictions on their flexibility.

4.15   Segment density distributions

A different problem concerns the time-average density of monomer units or segments in a volume element in the vicinity of a polymer chain. In section 3.17 we assumed that the distribution was uniform throughout the domain of the polymer molecule, but this is a gross oversimplification.

Let us suppose that the time-average number of monomer units (or segments) within a small volume element $\mathrm{d}v$ situated at a distance s from the centre of mass of the polymer is $\rho(s) \, \mathrm{d}v$. It has not been possible to derive an exact expression for $\rho(s)$ even for chains of infinite length. Nevertheless an approximate expression is available for unperturbed random-flight polymers which takes the form

$$\rho(s) = A \, \mathrm{e}^{-Cs^2} \tag{4.27}$$

where A and C are constants. The derivation of this expression is arduous, but it is based on the results discussed in the previous section. The constants A and C may be evaluated by noting that the integral of $\rho(s)$ over all space must equal σ, the total number of segments, and that the integral of $s^2\rho(s)$ over all space may be equated to $\langle s^2 \rangle_0$, the radius of gyration of the random coil. After performing these integrations we see that

$$\rho(s) = \sigma \left(\frac{3}{2\pi\langle s^2 \rangle_0} \right)^{3/2} \mathrm{e}^{-(3s^2/2\langle s^2 \rangle_0)} \tag{4.28}$$

This expression implies that the segment density is greatest at the centre of mass of the domain and equal to $\sigma(3/2\pi\langle s^2 \rangle_0)^{3/2}$. In section 4.10 we have seen that, for long polymer chains, $\langle s^2 \rangle_0$ may be written as $\sigma \beta^2 / 6$. We thus see that the segment density at the centre of the domain is inversely proportional to $\sqrt{\sigma}$.

Equation (4.28) also implies that the distribution about the centre of mass is spherically symmetrical. More accurate treatments of the problem, though still inexact, have suggested that this is not so, and the shape of the domain is better represented by a prolate ellipsoid of revolution.

A further consequence of equation (4.28) is that the fraction of space occupied by polymer segments within the domain is low. Thus at the centre of mass of a polymethylene molecule of carbon–carbon bond length 0.11 nm, degree of polymerisation 1000 and characteristic ratio 6.7; the density of segments at the centre of the domain is 6.6 segments per nm^3. Taking the excluded volume of a methylene group as $0.023 \ nm^3$, the fraction of space occupied is 15%.

So far we have only considered the density distribution function that is expected to pertain for the unperturbed polymer, that is at theta conditions. In good solvents the domain expands by a linear factor α (as defined in section 4.9). The result is that the density of segments decreases still further. In chapter 6 we shall argue that the centre of the domain expands more than the periphery. This results in the distribution becoming non-Gaussian in good solvents.

Problems

4.1 The molecular weight and radius of gyration of a monodisperse preparation of polymethylene were measured in a theta solvent and found to be 22 400 and 6.5 nm respectively. Calculate the mean square end-to-end distance, the characteristic ratio and the persistence length. The carbon–carbon bond length may be taken to be 0.153 nm.

4.2 Sketch plots of the characteristic ratio of a freely rotating coil as a function of the degree of polymerisation. Assume bond angles of $90°$ and $120°$.

4.3 A protein consists of eight identical spherical subunits of radius 5 nm, in contact and with their centres at the corners of a cube. Estimate a value for the radius of gyration. How does the fact that the subunits are spherical enter into your result?

4.4 The protein collagen is believed to have the shape of a cylindrical rod. Its molecular weight has been measured and found to be 345 000 and its radius of gyration 86 nm. Calculate the length and radius of the rod assuming that the density of the particle is $1.43 \ g/cm^3$.

4.5 The persistence length of double-stranded DNA has been reported to be 42 nm. Calculate the root mean square end-to-end distance for DNA molecules of molecular weight 10^7, 10^6 and 10^4. Assume that they behave like worm-like coils; that the axial separation between successive base pairs is 0.34 nm; and that the mean molecular weight of a base pair is 700.

4.6 Estimate the volume fraction of polymer at the centre of the domain for a sample of polyuridylic acid in a theta solvent given that the degree of polymerisation is 1000; the radius of gyration is 125 nm and the partial molar volume of a monomer unit is 177 cm^3/mol. What would the volume fraction be if the degree of polymerisation were 10^6?

4.7 A globular protein associates in a certain solvent. When this happens, the molecular weight and radius of gyration both increase by the same factor. Suggest a model to describe the manner in which the monomers are arranged in the polymer.

4.8 Polyuridylic acid has a characteristic ratio of 17.6 in a theta solvent. It can associate with a molecule of a polyadenylic acid to form a two-stranded helical complex resembling DNA in which the axial distance between adjacent base pairs is 0.34 nm. By what factor will the radius of gyration of the rigid helix exceed that of a polyuridylic acid contained in it. Assume that the degree of polymerisation of the polyuridylic acid is 100 and that the average bond length is 0.23 nm.

5　Helix–coil transitions and denaturation

We have seen in chapter 4 that, at low temperatures or in appropriate solvents, certain polymers adopt a helical conformation in which the dihedral angles associated with any given monomer unit have the same values as those associated with any other unit. As the temperature is raised or the nature of the solvent is altered, there is an ever increasing probability that these angles adopt values incompatible with the helix and of kinks appearing. As this happens the helix is said to undergo a *helix–coil transition*, or 'melt', till in the limit the angles are distributed among their possible minima at random and the polymer becomes a random coil.

As we shall see in more detail below, such a helix–coil transition may be *cooperative* or *non-cooperative*. In the latter case the probability of a kink occurring at a particular position is independent of whether kinks have appeared elsewhere in the chain; under these circumstances the helix 'melts' gradually. In the former cooperative case, the probability of the occurrence of a kink at one position is strongly coupled to the probability of there being a kink in neighbouring positions. This may result in the helix–coil transition taking on an all or none character, so that any given chain will be either completely helical or in the random-coil state, according to the temperature (or other conditions). In this chapter we will explore some of the general features of this effect.

We have also seen in chapter 4 that certain polymers, globular proteins in particular, can adopt a compact folded conformation which constitutes their tertiary structure. These too may revert to the random-coil state or denature when the temperature or solvent is changed. Such denaturation processes are also characteristically cooperative, and in this chapter we shall describe general features of denaturation as well.

5.1　Partially helical states

We shall suppose that a polymer molecule that is capable of adopting a helical conformation is composed of a string of units, each of which may be in a configuration corresponding to the helical conformation or in one or more alternative configurations which are inconsistent with the helix. Each such unit might correspond to a monomer unit which would be a natural choice in the case of a polypeptide, or it might correspond to a base pair in double-helical polynucleotides, or even to a simple dihedral angle. The exact nature of the relation between a helix unit and the structural features of the polymer will depend on the polymer under consideration; there may even be several different sorts of helix unit, but for simplicity we shall assume that there is only one. Let us denote a unit in the helix configuration by the symbol h and a unit in a non-helical configuration by the symbol c.

We now suppose that any given state of the polymer chain has some of its units in the h-state and some in the c-state. The state of the chain can then be denoted by a sequence of the two symbols h and c (e.g. hhchcchhchchc ...), n of them altogether. A sequence consisting entirely of h's would correspond to the fully helical conformation, and a sequence entirely composed of c's to the random-coil state. In general a sequence of both h's and c's would correspond to a partially helical state.

It should be noted that the specification of the state of the chain by a sequence of symbols in this manner does not define a conformation of the chain in the sense discussed in chapter 4, since a unit in a c-state may correspond to several different values of the dihedral angle or angles associated with the helix unit. The sequence only specifies a state of the chain as relevant to the helix–coil transition.

For a chain of n units, each in either an h-state or a c-state, there are clearly 2^n possible states of the chain as a whole. Each of these states will have a certain energy and a certain entropy associated with it.

If there are x h-units in the sequence, there are clearly $n - x$ c-units. We also note that any given sequence of h's and c's consists of runs of consecutive h-units (consisting of one or more h-units) punctuated with runs of consecutive c-units (of one or more c-units). We shall suppose that there are y runs of h-units or y *helical regions*. Each helical region (excepting possibly the first) starts with a c-unit immediately followed by a h-unit.

The helix content of a particular partially helical state of the chain we define as the ratio x/n and denote by θ. We shall be interested in the average value of the helix content, $\langle \theta \rangle$, over an ensemble of chains. If all the different states are in equilibrium this will be the same as the time-average helix content.

5.2   The partition function for partially helical chains

Let us suppose that a solution of a polymer contains all possible partially helical states in equilibrium. Let us also suppose that we have allocated a number to each of the states reserving 1 for the non-helical random coil (cccc ... c). The proportions of all the possible states that are present at equilibrium will be determined by a set of equilibrium constants so that, for any state numbered i, we have $c_i = c_1 K_i$, where c_i is the concentration of chains in state i. We then see that the proportion of chains in state i is given by p_i in

$$p_i = K_i \bigg/ \sum_{i=1}^{2^n} K_i \qquad (5.1)$$

The equilibrium constants, K_i, can be expressed in terms of the standard molar enthalpy and entropy differences between the random-coil state (state 1) and state i by means of the van't Hoff relation, so that $-RT \ln K_i = \Delta H_i^{\ominus} - T\Delta S_i^{\ominus}$. We then see from equation (5.1) that

$$p_i = e^{-(\Delta H_i^{\ominus} - T\Delta S_i^{\ominus})/RT} \bigg/ \sum_{i=1}^{2^n} e^{-(\Delta H_i^{\ominus} - T\Delta S_i^{\ominus})/RT} \qquad (5.2)$$

The sum in the denominator of equation (5.2) is defined to be the partition function, Q, of the ensemble of states. It should be noted that there is one term in the partition function to correspond to each of the 2^n possible states.

Let us suppose that state i contains x_i h-units and y_i helical regions. It is then clear that the average number of h-units per chain, $\langle x \rangle$, and the average number of helical regions per chain, $\langle y \rangle$, are given by $\sum_1^{2^n} p_i x_i$ and $\sum_1^{2^n} p_i y_i$ respectively. The helical content, $\langle \theta \rangle$, is then given by $\langle x \rangle / n$. In order to proceed further we need to evaluate $\Delta H_i^{\ominus}$ and $\Delta S_i^{\ominus}$ in terms of x_i and y_i. To do this we shall introduce the simplifying assumption that the interconversion of one state to another proceeds without any change in volume; this enables us to put the enthalpy change $\Delta H_i^{\ominus}$ equal to the energy change $\Delta U_i^{\ominus}$. This assumption may not be entirely warranted in practice if the formation of a helical unit from a c-unit is accompanied by a change in its interaction with the solvent, but is justified in the context of the schematic account of helix-coil transitions that we are presenting.

5.3 The energy and entropy of partially helical states

Let us consider the process of forming a state of the chain characterised by x_i h-units and y_i helical regions from the random-coil state ($x_1 = y_1 = 0$). We first suppose that the formation of each h-unit from a c-unit is accompanied by a change in the energy of the chain equal to ΔU_h. There is thus a contribution to the total energy change equal to $x_i \Delta U_h$. This contribution does not give the total energy change in general because the energy associated with an h-unit depends on the state of its neighbours along the chain.

In this schematic account we shall assume that such neighbour interactions only extend as far as nearest neighbours along the chain. It must be realised however that for real polymers that exhibit helix–coil transitions, such as polypeptides and DNA, this is not strictly true. We only point out that the theory may be elaborated with greater or lesser difficulty to take into account the possibility of more distant interactions.

In terms of this simplification, we now suppose that for each c-unit followed by an h-unit, there is an extra energy increment, ΔU_{ch} and similarly define increments ΔU_{hh} and ΔU_{hc}. We then see that the total energy increment associated with the formation of a helical region containing x h-units is $x\Delta U_h + \Delta U_{ch} + (x - 1)\Delta U_{hh} + \Delta U_{hc}$. This may be conveniently written in the form $\Delta U_j + x\Delta U_p$ where $\Delta U_p = \Delta U_{hh} + \Delta U_h$ and $\Delta U_j = \Delta U_{ch} + \Delta U_{hc} - \Delta U_{hh}$.

We then see that for a partially helical state of the chain, with a total of x_i h-units and y_i helical regions, the total energy change which occurs when it is formed from the random-coil state is

$$\Delta U_i^{\ominus} = x_i \Delta U_p + y_i \Delta U_j \tag{5.3}$$

In writing equation (5.3) we have overlooked one small difficulty, in that we have glossed over what happens if the first unit in the chain is in an h-state or if the last is. Equation (5.3) is valid on the assumption that a helical region at the start of the chain (and hence not preceded by a c-unit) has the same energy as if it were preceded by one or more c-units, and similarly that a helical region at the end has the same energy as it would if it were followed by a c-unit.

Exactly similar considerations apply to the entropy change so that

$$\Delta S_i^{\ominus} = x_i \Delta S_p + y_i \Delta S_j \tag{5.4}$$

We shall now define the *helix initiation constant*, σ, as $e^{-(\Delta U_j - T\Delta S_j)/RT}$ and the *helix propagation constant*, s, as $e^{-(\Delta U_p - T\Delta S_p)/RT}$. It then follows from equations (5.3) and (5.4) that

$$K_i = s^{x_i}\sigma^{y_i} \tag{5.5}$$

and that the partition function, Q, may be written in the form

$$Q = \sum_{i=1}^{2^n} s^{x_i}\sigma^{y_i} \tag{5.6}$$

The significance of the initiation constant derives from the argument presented above which suggests that an extra increment of energy and entropy may be required to initiate a new helical region above and beyond that necessary to form the h-units themselves. The formation of the h-units as such is accommodated in the propagation constant. If the initiation constant is very small, partially helical states of the chain with more than one helical region ($y_i > 1$) will, as indicated by equation (5.5), be associated with a very small equilibrium constant and hence will occur but rarely in the equilibrium mixture of states. This will be so, no matter how many h-units there are. A small value of σ requires that the free energy of formation of a helical region, $\Delta U_j - T\Delta S_j$ be large and positive. It is often assumed that σ is independent of temperature and hence that the free energy of initiation is entropic in origin. We shall discuss later the physical origin of this free energy of helix initiation.

We are now in a position to write down expressions for $\langle x \rangle$ and $\langle y \rangle$:

$$\langle x \rangle = \sum_i p_i x_i = \sum_i x_i s^{x_i} \sigma^{y_i} \Big/ \sum_i s^{x_i} \sigma^{y_i} = \frac{\mathrm{d}\ln Q}{\mathrm{d}\ln s} \tag{5.7}$$

$$\langle y \rangle = \sum_i p_i y_i = \sum_i y_i s^{x_i} \sigma^{y_i} \Big/ \sum_i s^{x_i} \sigma^{y_i} = \frac{\mathrm{d}\ln Q}{\mathrm{d}\ln \sigma} \tag{5.8}$$

We thus see that the evaluations of $\langle x \rangle$ and $\langle y \rangle$ require an expression for the partition function, Q. This is seen from equation (5.5) to be a polynomial in s and σ, with one term for every possible state.

5.4 The evaluation of the partition function

The partition function, Q_n, for a chain of n units is the sum of 2^n terms of the form $s^x \sigma^y$ with x and y taking all possible values consistent with n. We reiterate that each term corresponds to a possible partially helical state of the chain. Let us begin by dividing these states into two classes A and B. The states in class A are those in which the last unit of the chain is in a c-state. There will be 2^{n-1} states in class A and let us suppose that the sum of the corresponding terms in the partition function is $Q_{n,c}$. The states in class B have an h-unit at the end; again there are 2^{n-1} members and we suppose that the sum of the terms in the partition function corresponding to them is $Q_{n,h}$. It is then clear that

$$Q_n = Q_{n,c} + Q_{n,h} \tag{5.9}$$

We now consider the problem of deriving expressions for the partial sums, $Q_{n+1,c}$ and $Q_{n+1,h}$, corresponding to chains with $n+1$ units, from expressions, $Q_{n,c}$ and $Q_{n,h}$, corresponding to chains with n units. There will be twice as many terms in both $Q_{n+1,c}$ and $Q_{n+1,h}$ than in $Q_{n,c}$ and $Q_{n,h}$ since the total number of states (2^{n+1} as opposed to 2^n) has doubled.

First consider one of the terms in $Q_{n,c}$. This corresponds to a state which terminates in a c-unit. If we add another unit to the end it may be either a c-unit or an h-unit. If we add a c-unit, no greater increment of energy or entropy would be required in the formation of the resulting state of the chain of $n+1$ units from the random-coil reference state than was required for the formation of the original state of n units. This implies that the equilibrium constants for the formation of the original and the new states are the same.

If, on the other hand, the extra unit added is in the h-state, an extra energy increment contingent on both forming a new h-unit and initiating a new helical region is required. This implies that the term in $Q_{n,c}$ must be multiplied by $s\sigma$ to give the equilibrium constant of the new state. In the former case the new term belongs to $Q_{n+1,c}$ and in the latter to $Q_{n+1,h}$. This argument holds for every term in $Q_{n,c}$ so that we see that half the terms in $Q_{n+1,c}$ are given by $Q_{n,c}$ and half those in $Q_{n+1,h}$ by $s\sigma Q_{n,c}$.

If we now consider the terms belonging to $Q_{n,h}$ we see that the new terms generated by adding a c-unit belong to $Q_{n+1,c}$ and are equal to those in $Q_{n,h}$, whereas those generated by adding an h-unit belong to $Q_{n+1,h}$ and are equal to those in $Q_{n,h}$ multiplied by s; for, although another h-unit is added, no new helical region is initiated.

Putting these observations together we readily see that

$$Q_{n+1,c} = Q_{n,c} + Q_{n,h}$$

$$Q_{n+1,h} = s\sigma Q_{n,c} + s Q_{n,h} \tag{5.10}$$

Since these arguments are slightly involved, we have attempted to delineate them and the derivation of equations (5.10) in a pictorial manner in figure 5.1.

Equations (5.10) may be used to generate the partial sums, $Q_{n+1,c}$ and $Q_{n+1,h}$ from $Q_{n,c}$ and $Q_{n,h}$ and hence, if $Q_{1,c}$ and $Q_{1,h}$ are given, to generate the expression for the partition function for chains containing any number of units. Consistent with our assumption that a state in which the first unit is in the h state behaves as if it were preceded by a c-unit, we may put $Q_{1,h} = s\sigma$ and $Q_{1,c} = 1$.

The derivation of an expression for Q_n in terms of s and σ is now a straightforward mechanical process, but for large n is extremely tedious. For this reason it is desirable to attempt to derive an explicit expression for Q_n. This can be accomplished in a straightforward manner using matrix algebra. The details of this derivation are given in appendix A for the interest of readers acquainted with this branch of mathematics. It may be noted in passing that similar mathematical stratagems are employed in the derivation of expressions for the mean square end-to-end distance of random-coil polymers in which there are nearest neighbour interactions as discussed in section 4.7.

The result of the application of the matrix method to equations (5.10) may be expressed by

$$Q_n = \frac{(\lambda_1^{n+1} - \lambda_2^{n+1}) - \lambda_1 \lambda_2 (\lambda_1^n - \lambda_2^n)}{\lambda_1 - \lambda_2} \tag{5.11}$$

where

$$\lambda_{1,2} = \frac{1+s}{2} \left\{ 1 \pm \left[1 - \frac{4s(1-\sigma)}{(1+s)^2} \right]^{1/2} \right\} \tag{5.12}$$

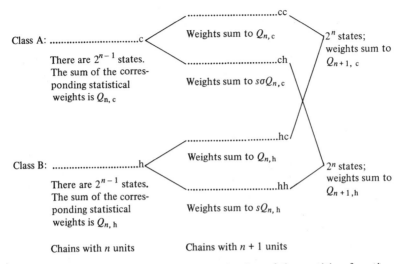

Figure 5.1. Illustration of the derivation of the partition function for a chain of $n + 1$ units from that for a chain of n units.

It may be seen from equation (5.12) that $\lambda_1 > \lambda_2$, so that for very long chains $\lambda_1^n \gg \lambda_2^n$ and in this limit the partition function, Q_n, reduces to λ_1^n. This approximation is valid for all values of s provided $n \gg 1/2\sigma^{1/2}$. The quantity of $1/2\sigma^{1/2}$ defines a characteristic length for a helical region in terms of the number of units (x) comprising it. If $n \approx 1/2\sigma^{1/2}$, a different approximation becomes valid in which all terms in the partition function corresponding to states of the chain with more than one helical region are dropped.

Equations (5.11) and (5.12) (or valid approximations thereto) allow the calculation of the partition function as a function of s, σ and n. They also allow, via equations (5.7) and (5.8), the calculation of expressions for $\langle x \rangle$ (and hence $\langle \theta \rangle = \langle x \rangle / n$) and $\langle y \rangle$, albeit with some effort. The results are complicated, and we shall not quote them explicitly.

Finally, we note as a point of terminology that the equilibrium constant for the formation of a particular state of the chain from the random-coil state which is represented by the corresponding term in the partition function may be referred to as the *statistical weight* of the state.

5.5 Cooperative and non-cooperative transitions

The meaning and significance of the results embodied in equations (5.11) and (5.12) and the corresponding expressions for $\langle x \rangle$ and $\langle y \rangle$ are not readily apparent. To bring out the important features we shall resort to graphical representations to show how relevant parameters vary with factors under experimental control.

First, let us note, we shall assume that the helix initiation constant, σ, is independent of temperature. This is equivalent to saying that the extra increment of free energy contingent on the formation of a helical region is entropic in nature. Although this assumption may not be strictly valid it appears to be consistent with experimental results for a variety of helix-coil transitions that have been investigated experimentally.

On the other hand we shall suppose that the propagation constant, s, is temperature dependent, so that s varies as the temperature is changed. It is convenient to define a temperature, T_m, as that at which $s = 1$. It then can be seen from the definition of s that $T_m = \Delta U_p/\Delta S_p$. We shall assume that ΔS_p is negative; this implies that the entropy of the chain decreases as c-units are changed into h-units. This is consistent with the notion that the helix is a rigid object with less conformational entropy than the random-coil state as discussed in section 3.19. We shall also assume that ΔU_p is negative, which is consistent with the idea that the helical form is stabilised by short range intramolecular interactions. In terms of these assumptions we then see that s decreases as the temperature is raised, being infinite at the absolute zero of temperature, unity at T_m and falling to a limiting value of $e^{\Delta S_p/R}$ at high temperature. It should be borne in mind that ΔS_p represents not only the decrease in conformational entropy when an h-unit is formed, but also any change in the entropy of the solvent which may occur in this process; it may also be that it varies with temperature. Similarly ΔU_p contains contributions from changes in the interaction with the solvent and may also vary with temperature. We shall assume that both parameters are independent of temperature for simplicity.

A method frequently employed in the study of helix-coil transitions is to determine $\langle\theta\rangle$, the helix content, as a function of temperature. Accordingly we display in figure 5.2 plots of $\langle\theta\rangle$ versus T (or more precisely $T - T_m$) for various values of σ. In these plots we have assumed that $T_m = 323$ K and that $\Delta S_p = -125$ J/K. This latter figure is characteristic of the helix-coil transitions of polynucleotides, but our qualitative conclusions would not be changed by other choices of value for ΔS_p. We have also assumed that n is large: 10^6. Before we proceed with a discussion of the salient features of these plots we may note that s can in principle be changed by altering the nature of the solvent as long as such changes alter ΔU_p or ΔS_p. Thus the abscissa in figure 5.2 could just as well represent some property of the solvent that varies, such as the concentration of a denaturing agent.

The first point to note from figure 5.2 is that $\langle\theta\rangle$ falls from a value close to unity at low temperatures to a value of $\frac{1}{2}$ at $T = T_m$ and then to a value close to zero at high temperatures. This conversion of the low temperature helical form to the high temperature random-coil form is the helix-coil transition. It is also sometimes convenient to refer to it as the 'melting' of the helix and plots such as those in figure 5.2 are sometimes called 'melting curves'. The temperature at which $\langle\theta\rangle = \frac{1}{2}$ is often called the *melting temperature*.

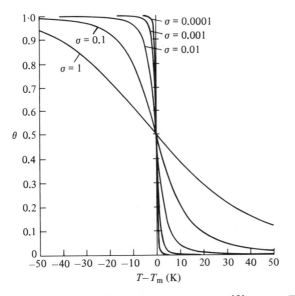

Figure 5.2. Plots of helical content, $\langle \theta \rangle$, versus $T - T_m$ for various values of the helix initiation constant, σ. T_m is the temperature at which the propagation constant becomes equal to unity. A value of 323 K was assumed for T_m and a value of -125 J/K for ΔS_p. The number of units in the chain was assumed to be 10^6.

We may also observe that, as the initiation constant, σ, decreases, the sigmoid curves become 'sharper'; that is to say that the temperature range between points where $\langle \theta \rangle = \frac{3}{4}$ and $\langle \theta \rangle = \frac{1}{4}$ becomes less and the slope of the curves at $\langle \theta \rangle = \frac{1}{2}$ becomes steeper. In the limit of $\sigma = 0$ we may show that $\langle \theta \rangle = 1$ for all temperatures below T_m and $\langle \theta \rangle = 0$ for all temperatures above T_m. In these limiting circumstances, the transition is said to be fully *cooperative* and takes on an 'all or none' character in that only the fully helical and fully random-coil forms have a finite probability of occurrence.

One may discern more clearly the origin of cooperativity by noting that the statistical weights of states of the chain with more than one helical region become very small if σ is small, since they contain higher powers of σ. This implies firstly that all the h-units tend to be coalesced into a single helical region and secondly that this can only melt by h-units at its ends converting to c-units.

The opposite behaviour is displayed when $\sigma = 1$. In this case the expression for $\langle \theta \rangle$ becomes simply $s/(1 + s)$ and the h-units convert to c-units quite independently. In this case the transition is non-cooperative. The smaller is σ, the more cooperative is the transition. In principle σ might be greater than unity, in which case the transition would be *anti-cooperative;* transitions of this nature are not observed in practice.

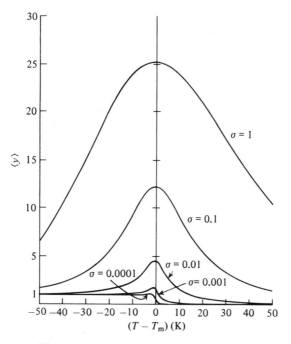

Figure 5.3. Plots of the mean number of helical regions per chain, $\langle y \rangle$, versus $T - T_{\mathrm{m}}$, for various values of σ. The parameters assumed were the same as in figure 5.2 except that the number of units in the chain was taken to be 100. Similar plots are obtained if it is assumed that there are 10^6 units but, in this case, details of the behaviour of the curves for different σ cannot be conveniently displayed on the same scale.

Further insight into the nature of cooperative transitions may be obtained from an examination of figure 5.3 in which we have plotted $\langle y \rangle$, the mean number of helical regions per chain, against the temperature for various values of σ. At low temperatures the chains are fully helical with but one helical region so that $\langle y \rangle$ is close to unity. In the case of cooperative transitions with low values of σ, $\langle y \rangle$ remains close to unity as the temperature is raised, but drops suddenly to zero at the melting temperature. On the other hand, for non-cooperative transitions or for transitions showing minimal cooperativity, $\langle y \rangle$ first increases as the temperature is raised as units in the body of the initial helical region revert randomly to c-units, so splitting the helical region into a larger and larger number of shorter regions. Only as the temperature is increased beyond the melting temperature does $\langle y \rangle$ decrease to zero as these short helical regions melt completely.

So far we have indicated that $\langle \theta \rangle = \frac{1}{2}$ when the temperature becomes equal to T_{m}. This is only strictly true in the limit of infinite n. For short chains with low n, $\langle\theta\rangle$ will become equal to $\frac{1}{2}$ at temperatures below the melting temperature for low values of σ. A rather rough approximation to the treatment of cooperative transitions suggests that $\langle\theta\rangle = \sigma s^n/(1 + \sigma s^n)$ when σ is small. $\langle\theta\rangle$ is then equal to $\frac{1}{2}$ when $s = \sigma^{-1/n}$ and this is significantly greater than unity (implying that the temperature is less than T_m) if both σ and n are small. The implication is that the free energy of helix initiation, $\Delta U_j - T\Delta S_j$, which is positive, decreases the overall stability of the cooperatively melting single helical region by decreasing the magnitude of its (negative) free energy of formation.

5.6 The mean square end-to-end distance of melting helices

The end-to-end distance of a helical polymer showing highly cooperative melting behaviour, and at sufficiently low temperatures that only one helical region is present, is given by the statistics of the worm-like coil discussed in section 4.12. In the melting region close to the melting temperature when the helix is breaking into shorter segments, a new and additional cause of chain flexibility is introduced since the c-units separating the helical regions allow kinking of the chain to occur. The chain then resembles a random-coil polymer, since it consists of a series of rod-like segments joined by sequences of flexible c-units.

Let us suppose that the projection of the length of an h-unit along the axis of the helix in which it participates is β_h. The square of the length of the complete helix of n units is then $n^2\beta_h^2$. Let us now suppose that the mean square end-to-end distance of a c-unit is β_c^2 and that two adjacent c-units are freely jointed. The mean square end-to-end distance of the random-coil state is then $n\beta_c^2$. Thus for $\langle\theta\rangle = 1$ (complete helicity) the characteristic ratio is $n(\beta_h/\beta_c)^2$, and for $\langle\theta\rangle = 0$ (zero helicity) it is unity. It is generally found that $\beta_h < \beta_c$. This implies that the characteristic ratio (or mean square end-to-end distance) may either increase or decrease as $\langle\theta\rangle$ changes from unity to zero and the helix-coil transition occurs. It will decrease for $n > (\beta_c/\beta_h)^2$ and increase if $n < (\beta_c/\beta_h)^2$. If the ratio β_c/β_h is equal to 5, the critical chain length at which the characteristic ratio remains unchanged after the completion of the melting process is given by $n = 25$.

For values of $\langle\theta\rangle$ intermediate between 1 and 0, the characteristic ratio may pass through a minimum. In the initial stages of the melting process, but few c-units are present; those that are present introduce kinks into the initially rigid rod. This necessarily leads to a collapse of the rigid structure and a concomitant decrease in the mean square end-to-end distance. Thus, in the initial stages of the melting process, the characteristic ratio decreases. It is only after a considerable number of c-units have been formed and if $\beta_c > \beta_h$, that their compensating effect may lead to a later increase in the characteristic ratio. Detailed theoretical treatments of the general effect have been developed but they are of too great a complexity to reproduce here.

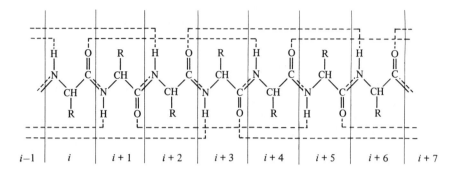

Figure 5.4. Schematic representation of the hydrogen-bonding scheme in α-helical polypeptides. The hydrogen bonds are indicated by dashed lines. The amino-acid residues (helical units) are enclosed within vertical lines.

5.7   Helix-coil transitions in real polymers

The analysis of helix–coil transitions presented above is the most simple that brings out the essential features of the phenomenon. Its application to real polymers requires the introduction of additional complexities, but the general features of the analysis remain unchanged.

Two types of polymer that manifest helix–coil transitions have been studied intensively: polypeptides which show a helix–coil transition between the α-helical conformation and the random-coil and the double-helical forms of naturally occurring polynucleotides, DNA and RNA and their double-helical and single-helical analogues.

The α-helical form of polypeptides is stabilised by hydrogen bonds between the carbonyl oxygen of one amino-acid residue and the amide hydrogen of the fourth succeeding residue along the chain, as schematically illustrated in figure 5.4. This in itself is sufficient to ensure that the transition is highly cooperative, for a nucleus of at least four amino residues must conform to the helical configuration before any stabilising hydrogen bond can be formed. To put the matter another way, three hydrogen-bond acceptor groups at one end of a helical region and three donor groups at the other end are unsatisfied; it is the absence of the (negative) free energy of formation of these missing hydrogen bonds that results in the positive free energy of helix initiation. An exact treatment of helix–coil transitions in polypeptides requires that account be taken of the reduced probability of occurrence of helical regions of less than four amino-acid residues. We shall not elaborate on the details of such treatments save to note that the theory outlined in previous sections of this chapter provides a good approximation to the more exact treatments. Values of σ of the order of 10^{-3} have been derived from the analysis of experimental melting curves.

Figure 5.5. Two partially melted states of a double-helical molecule. State I has two closed loops, each with the same number of dissociated base pairs. State II has a single closed loop with twice the number of dissociated base pairs as in either loop of structure I.

The helix–coil transition in double-helical polynucleotides presents even greater complexities. The unit of helicity in this case may be taken to be two nucleotide residues in the two opposing polymer chains. The heterocyclic bases of the two nucleotides are hydrogen bonded together in the helix to form a base pair, and these base pairs are stacked on top of each other in a helical region. The double helix is thus stabilised partly by hydrogen bonds between the bases forming the base pairs and partly by secondary interactions between adjacent base pairs in the stack. A base pair in the centre of a helical region thus interacts with the two base pairs on either side of it, but the two base pairs at the two ends of a helical region can only indulge in one such stacking interaction. It is the absence of these two putative interactions that is partly at the root of the cooperativity shown in the helix–coil transitions manifest in double-helical polynucleotides. An additional complexity arises from the fact that a sequence of c-units (in which the base pairs are dissociated) separating two helical regions constitutes a closed loop. The entropy associated with this closed loop is not proportional to the number of nucleotide residues, j, within it but to $\ln j^{3/2}$ as discussed in section 3.19. This implies that a state of the chain with several small non-helical regions has a lower statistical weight than one with the same number of base pairs and fewer non-helical regions. Since the helical and non-helical regions alternate, this also implies that the helical regions tend to coalesce. This is the second and major reason for the cooperativity of double-helical polynucleotides. To see this argument more clearly, consider the two states illustrated in figure 5.5. In state I, the two non-helical regions, each containing j dissociated base pairs, will each contribute a factor $(2j)^{-3/2}$ to the statistical weight. In state II there is a single non-helical region of $2j$ dissociated pairs which contributes a factor $(4j)^{-3/2}$. The ratio of the two statistical weights is then clearly $(2j)^{-3/2} \times (2j)^{-3/2}/(4j)^{-3/2} = j^{-3/2}$. Since j may be of the order of hundreds, it is clear that state II is more stable (higher statistical weight) than state I. Values of the helix initiation constant, σ, of the order of 10^{-4} have been determined from experimental melting curves.

A further complexity in the case of naturally occurring double-helical DNA and RNA arises from the differential stability of A.T (or A.U in the case of

RNA) and G.C base pairs. This means that the melting temperature (at which $\langle \theta \rangle = \frac{1}{2}$) depends on the relative proportion of G.C pairs. It also means that the melting may take place in a complicated non-random manner with regions of the double helix that are rich in A.T pairs melting at a lower temperature than G.C rich regions. In such circumstances the melting curve may show a series of steps, each step corresponding to the melting of such regions of different stability. Such structured melting curves have been observed for homogeneous DNA preparations of moderate molecular weight derived from certain viruses. In high-molecular-weight DNA, for which a large number of such steps might be anticipated, the individual steps are not resolved and the overall melting curve merely appears to be broadened beyond the limits set by the operative value of σ.

It is clear from these considerations that the details of the helix–coil transition for DNA depends on the sequence of base pairs along the double helix, and a knowledge of this sequence is required for the theoretical derivation of the form of the melting curve. Such sequences are available, and it is possible, with the aid of high-speed digital computers and estimates of the various thermodynamic parameters involved, to predict in numerical terms the form of the melting curve. Such theoretical curves are found to be in reasonable agreement with experiment, in vindication of the theory.

Finally, it may be noted that synthetic polynucleotides such as polyadenylic acid or polycytidylic acid also undergo a helix–coil transition, being helical and rod-like at low temperatures and random coil at high temperatures. In these cases (except in acidic solutions) a single rather than a double helix is formed. The h-units are stabilised by stacking interactions between adjacent bases, but the melting curves are broad, and the melting takes place over a considerable temperature range. This is in contradistinction to the sharp melting curves observed with double-helical polynucleotides. Such helix–coil transitions are essentially non-cooperative.

5.8   The denaturation and renaturation of globular proteins

If a solution of a globular protein is heated, or if increasing amounts of a denaturing agent (such as urea) are added, the protein denatures; the compact and highly determined tertiary structure is unfolded and the polymer reverts to the random-coil state. It might be thought, by analogy with helix–coil transitions, that this process of denaturation would occur via a large number of partially folded intermediates. However, much experimental work has indicated that denaturation generally proceeds in a highly cooperative manner via but a small number of intermediate states. In this respect denaturation resembles the melting of a three-dimensional crystal rather than that of a one-dimensional helix. There are no partially ordered forms of a crystal which occur as it melts; the constituent molecules are either organised in the crystal lattice or distributed at random in the melt.

Various reasons have been advanced to account for the cooperative nature of protein denaturation. One of the most compelling is based on the hypothesis that the compact folded tertiary structure is unique so that alternative compact intermediates with slightly higher energy do not exist to act as intermediates in denaturation. We have also seen that the tertiary structure is stabilised in part by long-range interactions between distant parts of the polypeptide chain which are in close proximity in the folded molecule. Such interactions may not take place unless both parts of the chain that are involved are folded in an appropriate manner, and such folding may not occur unless other interactions are present. In this way the tertiary structure may resemble a house of cards: a disturbance of any part of it leads to the collapse of the whole structure.

Finally we may note that the analogy between a helical polymer and a globular protein is very weak. It is true that the structure of both may be disrupted by heat or denaturing agents, but, in the partially helical forms of the former, there are a large number of almost equivalent states differing only in the relative number of c-states and h-states and the manner these are arranged in helical regions; this feature is not discernible in many globular proteins.

So far we have discussed denaturation as if it were an irreversible process. If this were truly the case, a thermodynamic analysis of denaturation would be out of place and it would follow that the native tertiary structure of a globular protein was not a thermodynamically stable state and that the long term existence of globular proteins in their native structure was due to kinetic factors rather than thermodynamic ones. However, it has been possible to bring about the renaturation of denatured proteins, with the implication that a true thermodynamic equilibrium can exist between the native and denatured forms.

This raises an interesting paradox. If the renaturation process proceeds without the formation of intermediate forms, renaturation can only occur if the polypeptide chain spontaneously arranges itself into the native structure by a trial and error process. However, the number of possible conformations of a polypeptide chain of, for example, 100 amino-acid residues is of the order of 2^{200}. If these were tried out at the rate of 10^6 per second, we would expect to have to wait for 10^{22} years before the correct conformation was found by chance. This period of time is far greater than the age of the universe, but proteins sometimes renature in a matter of seconds. One possible resolution of this paradox is to suppose that intermediates are indeed formed in the renaturation and denaturation reactions which direct the pathway of these processes and that these are but short lived and present in indetectable amounts. We can only conclude that an adequate theory of protein renaturation and denaturation still remains to be developed.

Problems

5.1 Polyadenylic acid appears to undergo a helix-coil transition on heating at pH 7.0. The transition was studied by monitoring its optical rotation as a function of temperature and by microcalorimetry. The enthalpy change on formation of the helical form per nucleotide residue was found to be -45 kJ/mol and the proportion of residues in the helical form was found to be 0.25 at $63.2°C$ and 0.5 at $41.1°C$. Do these results support the hypothesis that the transition is non-cooperative?

5.2 The T_m of a sample of DNA was found to be $69°C$ in 0.01 M NaCl and $87°C$ in 0.1 M NaCl. In either solvent the enthalpy of the helix-coil transition was found to be -9.2 kJ/mole of base pairs. Calculate the entropy of the transition in the two solvents.

5.3* Write an expression for the partition function for a tetramer undergoing a helix-coil transition in terms of the helix propagation and initiating constants. Hence derive an expression for the helix content.

5.4 Given the expression obtained in problem 5.3, sketch a graph of the helix content at the temperature for which the propagation constant is equal to unity versus the initiation constant.

5.5 The enthalpy of denaturation at $45°C$ of a protein was found to be 370 kJ/mol from calorimetric studies. At the same temperature the protein was found to be half denatured. Estimate the entropy of denaturation assuming the denaturation is an all or none phenomenon. There are 124 amino acids in the single polypeptide chain. Calculate the entropy of denaturation per amino acid. Comment on your result.

6 Gels and polymer networks

We have seen in chapter 1 that a gel or polymer network consists of a series of polymer chains which are cross linked to form a single giant molecule that resembles a three-dimensional net. Since the individual polymer chains are joined together so that they cannot move freely with respect to each other as they can in free solution, the gel resembles a solid more than a liquid in that it exhibits an elastic deformation in response to an applied stress.

The deformation produced by such a stress is transmitted to the individual polymer chains with the result that their conformational entropy is changed. This as we shall see below gives rise to an elastic force which opposes the applied stress. In this way we can account for the elasticity displayed by cross-linked networks such as cured rubber.

The polymer network may have an open structure such that a considerable quantity of solvent may permeate its interstices. We have discussed in chapter 3 how the mixing of polymer chains with solvent results in a reduction in the free energy of the system. In the same way a network may absorb solvent with a concomitant reduction in free energy. For this reason a gel immersed in solvent may tend to swell by imbibing solvent. As it does so the dimensions of the network chains alter and the conformational entropy of the polymer chains alters as described above. This results in an increase in the free energy. The net result is that the gel swells till the two effects, one tending to increase the free energy and the other tending to decrease it, balance. We discuss this effect also in this chapter.

Finally, we may note that a single isolated polymer chain resembles a minute gel in that an expansion of the coiled up chain leads to a decrease in conformational entropy and the concomitant increase in the amount of solvent within its domain results in an increase. We shall see that the nature of the balance struck between these tendencies accounts for the chain expansion factor introduced in section 4.9.

6.1   The structure of the gel network

If N_2 polymer chains are joined together by cross links to form a single molecule, $N_2 - 1$ cross links are required. Any additional cross links

115

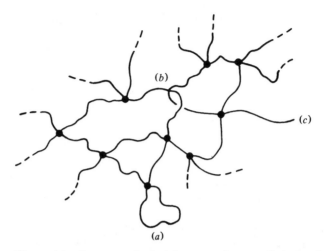

Figure 6.1. Some topological features of a cross-linked polymer network. (*a*) A portion of a chain may be looped back on itself and
only attached to the network at one point. The portion of the chain
in the loop is not an active chain since it cannot be under a tension
transmitted from the rest of the network. (*b*) Two interlocking loops
forming an entanglement; the entanglement can act as an additional
cross link. (*c*) A free end of a chain is attached at only one point to
the rest of the network and hence is not an active chain.
In the diagram, cross links are represented as dots.

introduced will then result in the formation of closed loops. Each closed loop
will consist of at least two cross links joined by two chains which we shall
call *active chains*. If N_c is the total number of cross links, the number of
active chains is $2(N_c - N_2 - 1)$. If, as is generally the case, $N_c \gg N_2$, this
reduces to $2N_c$ active chains as an approximation. If the gel as a whole is
distorted, each of the active chains is distorted in that the distance between
the two cross links at its ends is altered.

This simple picture of a cross-linked gel, we must note, is somewhat oversimplified. In the first place a cross link *may* serve only to loop a chain back
on itself (as indicated by point (*a*) in figure 6.1). The polymer chain in such
a loop, because it is only joined to the rest of the network at one point,
should not be counted as an active chain. On the other hand, two closed
loops may be interlocked (as indicated by point (*b*) in figure 6.1); such an
entanglement may act as an extra cross link and generate two more active
chains. An analysis of such complications is extremely difficult and so we
shall, for simplicity, assume that the number of active chains is twice the
number of cross links.

It is worth noting that entanglements of the kind noted above may exist

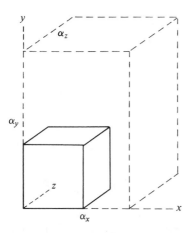

Figure 6.2. The deformation of a unit cube of gel (solid figure) into a rectangular parallelepipedon (dashed figure).

in concentrated solution of polymers containing no cross links. To be sure these will slowly unravel as a result of the random Brownian motion of the chains, but only to reform. It is partly for this reason that solutions of some polymers may exhibit elastic properties when subject to sudden stresses.

6.2 The free energy of deformation of a gel

We wish to derive an expression for the change in free energy accompanying the deformation of a gel. For simplicity we shall consider a portion of a gel in the form of a cube of unit volume. We now suppose that this cube is deformed into a parallelepipedon with sides equal to α_x, α_y and α_z as indicated in figure 6.2. We also suppose that the components of the end-to-end displacement of any active chain in directions parallel to the sides of the cube are changed from x, y, z to $\alpha_x x$, $\alpha_y y$, $\alpha_z z$. This change in the end-to-end distances of the chains gives rise to a change in the entropy of the whole gel which we denote by $\Delta S'(\alpha x, \alpha y, \alpha z)$.

For the time being we shall assume that the number of monomer units or segments in each active chain is the same and given by σ. This artificial constraint on the structure of the gel will be removed later. We shall also assume that the end-to-end vectors of the active chains are distributed in magnitude and direction in the same manner as they would be if they were not linked to form the gel. Thus the number of active chains such that the components of the end-to-end vector lies in the ranges x to $x + dx$, y to $y + dy$ and z to $z + dz$ is equal to $2N_c W(x, y, z) \, dx \, dy \, dz$ where $W(x, y, z)$ is given by equation (4.24) to a reasonable approximation and providing the end-to-end distance is not too great. Thus

$$2N_c W(x, y, z) = \frac{2N_c b^3}{\pi^{3/2}} e^{-b^2(x^2 + y^2 + z^2)} \tag{6.1}$$

Here we have replaced $3/2\langle r^2 \rangle_0$ in equation (4.24) by b^2 for simplicity.

Now we have also seen in section 3.19 that when one chain is deformed so that the components of its end-to-end vector, originally in the ranges x to $x + dx$, y to $y + dy$ and z to $z + dz$, are changed to lie in the ranges $\alpha_x x$ to $\alpha_x x + \alpha_x \, dx$, $\alpha_y y$ to $\alpha_y y + \alpha_y \, dy$ and $\alpha_z z$ to $\alpha_z z + \alpha_z \, dz$, the conformational entropy of the chain changes by an amount given by equation (3.53) as $\delta S'$ in

$$\delta S' = -kb^2 \left[(\alpha_x^2 - 1) x^2 + (\alpha_y^2 - 1) y^2 + (\alpha_z^2 - 1) z^2 \right] + k \ln (\alpha_x \alpha_y \alpha_z) \tag{6.2}$$

It follows that the total change in entropy due to the deformation of all chains with end-to-end vector components in these ranges is given by $2N_c W(x, y, z) \, \delta S' \, dx \, dy \, dz$, and that the total entropy change for all $2N_c$ chains is obtained by integrating the resulting expression over all values of x, y and z. This integration is easily performed with the result that

$$\Delta S'(\alpha_x, \alpha_y, \alpha_z) = N_c k [2 \ln (\alpha_x \alpha_y \alpha_z) - \alpha_x^2 - \alpha_y^2 - \alpha_z^2 + 3] \tag{6.3}$$

It may be observed that this final expression for the entropy change is independent of the constant b and hence independent of the number of segments in the active chain. It is, however, proportional to the number of chains. This implies that, if we took a total of $2N_c$ polydisperse chains, the total entropy change on deformation would still be given by equation (6.3). Thus our assumption that the chains each contained the same number of segments is of no consequence and need not be made.

This derivation of equation (6.3) has tacitly ignored the fact that the chains are joined together in the network; thus equation (6.3) would be valid for any solution of polymer chains provided each was deformed in the manner described. If we allow for the fact that the chains are joined together, an additional contribution to the entropy change arises. This may be evaluated as follows.

We imagine the cross links to be formed in the following manner. First, we decide in advance which of the $4N_c$ chain ends are to be joined at each of the N_c cross links. We then suppose that each cross link is formed when each of the four selected chain ends come together by chance in a small volume element δV without any disturbance of the end-to-end distribution (which thus remains in the form given by equation (4.24)). The proportion of the total number of configurations of the solution with any four ends within δV is proportional to $\delta V/V$, where V is the total volume of the solution (in our case $V = 1$). Thus the proportion of the total number of configurations such that each of the N_c sets of four ends are in their respective volume element is proportional to $(\delta V/V)^{N_c}$. This implies that the formation of the $2N_c$ cross links in this manner is accompanied by a decrease in the entropy (owing to the fact that the chain ends are no longer free to move about) equal to $k \ln (\delta V/V)^{N_c} + C$, where C is some constant. We now suppose that the gel is formed in a different manner by deforming the chains first and then forming the cross links. After deformation the volume of the solution is $\alpha_x \alpha_y \alpha_z V$ instead of V, so that the entropy decrease concomitant on forming the cross links is $k \ln (\delta V/\alpha_x \alpha_y \alpha_z V)^{N_c} + C$. Since the entropy change, as given by equation (6.3), which is concomitant on deforming the chains, must be the same whether the deformation takes place either before or after the formation of the cross links, the difference between $k \ln (\delta V/V)^{N_c}$ and $k \ln (\delta V/\alpha_x \alpha_y \alpha_z V)^{N_c}$ must represent the additional contribution to the entropy of deformation, $\Delta S''(\alpha_x, \alpha_y, \alpha_z)$, which we are seeking. Clearly then

$$\Delta S''(\alpha_x, \alpha_y, \alpha_z) = N_c k \ln (\alpha_x \alpha_y \alpha_z) \tag{6.4}$$

and the total entropy of deformation is equal to $\Delta S' + \Delta S''$ so that

$$\Delta S(\alpha_x, \alpha_y, \alpha_z) = c_c R [\ln (\alpha_x \alpha_y \alpha_z) - \alpha_x^2 - \alpha_y^2 - \alpha_z^2 + 3] \tag{6.5}$$

where c_c is the concentration of cross links ($N_c k = c_c R$).

We should expect that equation (6.5) is only valid for moderate values of the expansion factors, α, since the Gaussian expression for the distribution function on which it is based is not valid for large end-to-end distances. We should also note that the concentration term c_c should in fact be derived from the number of cross links or entanglements which are effective in producing active chains rather than from the number of chemical cross links introduced, and hence is never known precisely.

Equation (6.5) gives the entropy of deformation of the gel. The free energy of deformation is given by $\Delta H(\alpha_x, \alpha_y, \alpha_z) - T\Delta S(\alpha_x, \alpha_y, \alpha_z)$, where $\Delta H(\alpha_x, \alpha_y, \alpha_z)$ is the enthalpy of deformation. If the chains deform with no change in their volume, this is equal to the energy of deformation. When a polymer chain is deformed, it may be that the relative number of dihedral angles of the polymer backbone that are in the different states of minimum energy changes. If this is the case the energy will change on deformation; this is clearly the case for large deformations in which, in the case of polymethylene chains for instance, all of the angles would be forced into the *trans* configuration at maximum extension. Nevertheless, for small deformations the redistribution of bond rotational states may be minimal so that the energy of deformation is small.

Finally, it should be noted that equation (6.5) can only be expected to hold at the theta temperature because the distribution function on which it is based only holds under such conditions.

6.3   Rubber elasticity

The account of gel deformation given in the last section readily yields a theory of the elasticity of rubber and similar cross-linked gels. Consider a piece of rubber of length L_0 in the unstretched state and of unit cross-sectional area. We suppose that a tension, F, is applied between the two ends causing it to stretch to a length, L, so that the deformation ratio in the x direction is $\alpha_x = L/L_0$. We further suppose that this deformation occurs at constant volume so that $\alpha_x \alpha_y \alpha_z = 1$. We may then put $\alpha_x = \alpha$ and $\alpha_y = \alpha_z = 1/\alpha^{1/2}$. Putting these values of the deformation ratios into equation (6.5), we see that the change of entropy per unit volume is given by

$$\Delta S = -c_c R (\alpha^2 + 2/\alpha - 3) \tag{6.6}$$

If we differentiate this expression with respect to the length, L, we obtain the rate of change of entropy per unit volume with length. Performing this operation, replacing $d\alpha/dL$ by $1/L_0$ and multiplying by L_0 we obtain the rate of change of entropy with length of the whole piece of rubber as

$$(\partial S/\partial L)_{T,V} = -2c_c R (\alpha - 1/\alpha^2) \tag{6.7}$$

We must now relate this quantity to the equilibrium tension, F, that produces this extension, α, by means of a thermodynamic argument. First we note that, at equilibrium, we have the following relation between small changes in the functions of state:

$$dU = T\,dS - P\,dV + F\,dL \tag{6.8}$$

a relation that follows directly from the first law of thermodynamics. From the definition of the Helmholtz free energy, $A(= U - TS)$, it then follows that

$$dA = -S\,dT - P\,dV + F\,dL \tag{6.9}$$

and hence that

$$F = (\partial A/\partial L)_{T,V} = (\partial U/\partial L)_{T,V} - T(\partial S/\partial L)_{T,V} \tag{6.10}$$

For moderate extensions we have argued that $(\partial U/\partial L)_{T,V}$ is small so that

$$F = -T(\partial S/\partial L)_{T,V} = 2c_c RT(\alpha - 1/\alpha^2) \tag{6.11}$$

This result implies that the tension is not proportional to the extension, α, and this prediction has been amply verified for rubber and other cross-linked gels. Experiment has also verified that equation (6.11) represents the correct relation between F and α as long as α is small. At high extensions, however, the relation no longer holds. The discrepancy at high extensions may be partly explained in terms of changes in the energy, which negate the assumption that $(\partial U/\partial L)_{T,V}$ is small and partly in terms of a tendency for the polymer chains that are then stretched out and roughly parallel to each other to form crystalline regions.

The theory outlined above suggests that the tension developed in stretched rubber is a result of the decrease in conformational entropy of the active chains. This is in contradistinction to the origin of the tension in other elastic solids (such as spring steel). In these the tension derives from the increase in energy of the material as the constituent atoms or molecules are displaced from their equilibrium positions as dictated by the secondary forces acting between them. This distinction has two interesting consequences. Firstly, that, if an elastic gel is stretched adiabatically, the temperature rises. Secondly that, if a gel under tension is heated, and the tension maintained constant, the length decreases. Both are at variance with the behaviour of steel.

For a reversible adiabatic change there is, by definition, no transfer of heat from the surroundings to the gel. This implies that the total entropy of the gel remains constant. However, we have seen that the conformational entropy decreases as the gel is stretched so that some other change resulting in an increase in entropy must occur at the same time if the gel is stretched adiabatically. This other change can only be a change in temperature. Since a rise in temperature is associated with an increase in entropy, we see that the temperature must rise. It is for analogous reasons that the temperature of a gas increases as it is compressed adiabatically. The decrease in entropy concomitant on the decrease in volume is compensated by an increase in temperature.

We have seen from equation (6.7) that $(\partial S/\partial L)_{T,V}$ is negative. Since it is possible to show that this quantity is equal to $(\partial F/\partial L)_{V,T}(\partial L/\partial T)_{V,F}$ the latter must also be negative. Since the length increases as the tension is increased, $(\partial F/\partial L)_{V,T}$ must be positive and hence $(\partial L/\partial T)_{V,F}$ must be negative. This shows as mentioned above that the length decreases as the temperature increases and the tension is maintained constant.

6.4   The swelling of gels

If the individual chains composing a cross-linked gel were not joined together, but enclosed in a semipermeable membrane immersed in an excess of solvent, the latter would diffuse into the bag until prevented by a rise in pressure or some other sort of compensation. From a thermodynamic point of view this is because the free energy of mixing the solvent with the solution is negative. If we now consider a gel in which the chains are cross linked, the surface of the gel acts as a semipermeable membrane, and solvent will diffuse in as before, provided the free energy of mixing is negative. There is, however, an important difference in that there is only one (giant) molecule of solute within the 'membrane'. This means that the cratic term in the free energy of mixing is absent and only the excess term need be considered. As the solvent diffuses into the gel, the latter swells and is deformed. At the same time the conformational entropy of the polymer chains decreases so that the concomitant part of the free energy of the gel increases. It is this increase in free energy that provides the compensating factor that prevents the gel from becoming infinitely swollen. A state of equilibrium is reached when the free energy decrease brought about by any further influx of solvent is exactly equal to the increase due to the further deformation. At this point the total free energy of the system is at a minimum.

To consider the effect in more quantitative terms let us suppose that, in the unswollen state at which the gel was formed, the weight concentration of polymer segments within the gel is ρ_2 and that its initial volume was V_0. Let us also suppose that, at equilibrium, n_1 moles of solvent have been absorbed leading to an isotropic swelling of the gel so that each dimension is altered in the same ratio, α. The volume of the swollen gel is clearly $V_0\alpha^3$. If V_1^* is the molar volume of the solvent and M_0 is the molar mass of a polymer segment, the volume fraction of polymer segments in the swollen gel is given by ϕ_2 in

$$\phi_2 = \rho_2 V_1^* / M_0 \alpha^3 \qquad (6.12)$$

If, in this swollen state, a further dn_1 moles of solvent are absorbed, a further change, dG, in the free energy of the gel will occur. This change in free energy will be the sum of two terms: $d\Delta G_m$ the change in the free energy of mixing of solvent with polymer network and dG_c the change in free energy concomitant on the further swelling of the gel. At equilibrium, dG must be zero, since the total free energy is at a minimum. This implies that $d\Delta G_m = -dG_c$ so that

$$d\Delta G_m/dn_1 = -dG_c/dn_1 \equiv -(dG_c/d\alpha)(d\alpha/dn_1) \qquad (6.13)$$

The quantity $d\Delta G_m/dn_1$ is clearly equal to the relative chemical potential of the solvent in the gel. We derive an expression for this by invoking the theory outlined in section 3.10. In particular we invoke equation (3.32). Before proceeding in this direction we note that there is only one polymer molecule in the gel: the cross-linked structure itself. We may ignore the fact that this molecule is cross linked and suppose that it behaves as a giant polymer chain containing a very large number of segments. This allows us to put $1/\sigma$ in equation (3.32) equal to zero. This also implies that the mole fraction of solvent within the gel is essentially unity so that the cratic part of the relative chemical potential of the solvent $(RT \ln x_1)$ is also essentially zero. We then see by putting $1/\sigma = 0$ in equation (3.32) and expanding the logarithmic term that to a first approximation

$$d\Delta G_m/dn_1 = RT\phi_2^2(\chi_1 - \tfrac{1}{2}) \qquad (6.14)$$

The factor $(\chi_1 - \tfrac{1}{2})$ may be replaced by $-\Psi(1 - \Theta/T)$ as usual.

To obtain an expression for $(d\alpha/dn_1)$ we note that the volume of dn_1 moles of solvent is $V_1^* dn_1$ and the absorption of this by the gel leads to a volume increase of $d(V_0\alpha^3)$ so that

$$d\alpha/dn_1 = V_1^*/3\alpha^2 V_0 \tag{6.15}$$

Finally, to obtain $dG_c/d\alpha$, we substitute $\alpha_x = \alpha_y = \alpha_z = \alpha$ into equation (6.5) and differentiate the result with respect to α. The result of this must be multiplied by V_0 to give the rate of change of entropy of the whole gel of volume V_0. Thus

$$dG_c/d\alpha = -3V_0 c_c RT(1/\alpha - 2\alpha) \tag{6.16}$$

Again, in deriving the free energy change from the entropy change we have assumed that the energy change is negligible.

If we now substitute from equations (6.14), (6.15) and (6.16) into (6.13) and eliminate ϕ_2 by using equation (6.12) we obtain

$$RT(\rho_2 V_1^*/M_0)^2 (\chi_1 - \tfrac{1}{2})/\alpha^6 = c_c V_1^* RT(1/\alpha^3 - 2/\alpha) \tag{6.17}$$

If the gel is highly swollen so that α is large this reduces to

$$\alpha^5 = (\rho_2/M_0)^2 (V_1^*/2c_c) (\tfrac{1}{2} - \chi_1) \tag{6.18}$$

For highly swollen gels, equation (6.18) requires that a plot of $\ln \alpha$ versus $\ln c_c$ be linear. This has been verified for several systems, but the slope of such plots is frequently different from the predicted value of 5. Equation (6.18) depends for its validity on both the theory of gel deformation and on the theory of concentrated polymer solutions as outlined in section 3.10 as well as on the approximation introduced in this section. It is thus not surprising that exact quantitative agreement with experiment is not obtained, but it is gratifying that a semi-quantitative agreement is observed. This suggests that the theory offered in the section is along the right lines.

Finally we may note that equation (6.18) is in any case only expected to be valid in good solvents for which α is large. The more accurate expression, equation (6.17), suggests that, in poor solvents, α may be less than unity. It is in fact observed that if a gel, originally formed in the presence of an excess of a good solvent, is immersed in a poor solvent, it will shrink or exhibit *syneresis*. This may be viewed as the equivalent of the precipitation of the polymer from a poor solvent. An aqueous polyacrylamide gel immersed in methanol behaves in this manner.

6.5   The expansion factor for flexible chains

We have seen in chapter 4 that the mean square end-to-end distance of a flexible polymer of sufficiently large molecular weight can be expressed as $\sigma\beta^2\alpha^2$. The expansion factor α is unity under theta conditions, but otherwise allows for the expansion or contraction of the coil in good or in poor solvents, so that $\langle r^2 \rangle / \langle r^2 \rangle_0 = \alpha^2$. We are now in a position to consider an approximate theory which relates α to the properties of the polymer chain.

We first recollect from section 3.17 that the presence of one polymer chain severely restricts the conformations available to another whose domain interpenetrates that of the first, but that the effect disappears at the theta temperature. This is because the repulsive interactions between the segments on one chain and those on another, which are due to the excluded-volume effect are compensated by attractive interactions. Exactly the same effects influence intramolecular interactions between one part of a chain and another. Since two parts of the same chain cannot occupy the same volume, the conformational states available to one part of the chain are restricted by the presence of other parts of the chain in the same domain, unless attractive interactions cancel the effect out. This leads to an expansion of the chain as discussed in section 4.9.

A quantitative theory of the effect can be constructed along the following lines. First we suppose that the polymer chain is contained within a small volume, its domain. In section 3.17 we made a similar supposition and supposed as an approximation that the density of segments within this domain was uniform, though a Gaussian distribution about the centre of mass would be a more realistic assumption. It turned out in section 3.17 that this approximation did not seriously affect the final result, and we shall assume here likewise that the segment density is uniform. Again it will turn out that the consequences of this assumption are not serious.

We now make the crucial step of likening this domain and the polymer coil therein to a tiny gel. This tiny gel may swell by imbibing more solvent with a concomitant decrease in the free energy of mixing, but with an increase in free energy contingent on the deformation of the polymer chain. Equilibrium is reached when the total free energy is at a minimum. It is true of course that there are no cross links in this tiny gel, but there is one active chain to be deformed, and this is all that is required.

The argument now proceeds along lines closely similar to those deployed in the previous section. The modifications required are as follows: first we replace the volume of the gel by the volume, V_d/N_A, of the domain; we next replace N_c, the number of cross links, by $\frac{1}{2}$ since there is only one active chain so that $c_c R$ is replaced by $\frac{1}{2}k$. We next note that the term $\Delta S''$ as given by equation (6.4) must not be included in the final expression for the entropy of deformation; this is because $\Delta S''$ represents an additional entropy change contingent on the fact that the active chains are cross linked; in the present situation there are no cross links so there is no counterpart to $\Delta S''$. With these modifications in mind we obtain the results

$$d\Delta G_m/dn_1 = RT\phi_2^2(\chi_1 - \tfrac{1}{2}) \tag{6.19}$$

$$d\alpha/dn_1 = N_A V_1^*/3\alpha^2 V_d \tag{6.20}$$

$$dG_c/d\alpha = -3kT(1/\alpha - \alpha) \tag{6.21}$$

which are analogous to equations (6.14), (6.15) and (6.16) respectively.

We now replace ϕ_2, the volume fraction of polymer in the domain, by $V_2^*/\alpha^3 V_d$, where V_2^* is the molar volume of the polymer chain, and find by invoking equation (6.13) that

$$\alpha^5 - \alpha^3 = (\tfrac{1}{2} - \chi_1) V_2^{*2}/V_1^* V_d = X \qquad (6.22)$$

The parameter X is identical to that introduced in section 3.17 and defined by equation (3.45).

If the more appropriate Gaussian distribution of segments within the domain is assumed, a more elaborate derivation is required, but an expression identical to equation (6.22) is obtained provided the radius of the domain is set equal to $(4\pi/3)^{1/6} \langle s^2 \rangle_0^{1/2}$, where $\langle s^2 \rangle_0^{1/2}$ is the unperturbed radius of gyration of the polymer chain. If we further replace $\tfrac{1}{2} - \chi_1$ by $\Psi(1 - \Theta/T)$ we find then that

$$\alpha^5 - \alpha^3 = 2C_M \Psi(1 - \Theta/T) M_2^{1/2} \qquad (6.23)$$

where M_2 is the molar mass of the polymer and C_M is a constant that depends only on the nature of the polymer but not on its molecular weight. It is given by

$$C_M = \tfrac{1}{2}(9/2\pi)^{3/2} (\bar{v}_2^2/N_A \overline{V}_1) (M_2/\langle r^2 \rangle_0)^{3/2} \qquad (6.24)$$

It should be noted that we have replaced $\langle s^2 \rangle_0$ by $\langle r^2 \rangle_0/6$ in accordance with equation (4.3) and that the ratio $M_2/\langle r^2 \rangle_0$ is independent of σ or the molecular weight of the polymer if this is large.

An important consequence of equation (6.23) is that the expansion factor becomes equal to unity at the theta temperature. Thus experimental values of $\langle r^2 \rangle$, which are to be compared to values calculated by the methods outlined in chapter 4, should be obtained with the polymer under theta conditions. Under these same conditions the second virial coefficient is zero.

At the theta point, $\alpha = 1$ so we see from the relation $\langle r^2 \rangle = \sigma\beta^2\alpha^2$ that $\langle r^2 \rangle$ is proportional to σ. At another extreme, when the polymer is highly expanded in a good solvent, α is proportional to $M^{0.1}$ so that $\langle r^2 \rangle$ is proportional to $\sigma^{1.2}$. These conclusions have been amply verified for a variety of polymers. It should be noted however that the numerical value of C_M as calculated from equation (6.24) does not agree, in general, with experimental values. Thus the agreement between the theory and experiment is only semiquantitative. This is not surprising considering the numerous assumptions and approximations involved, as we have discussed in the previous chapter. Nevertheless the theory does enable us to define the circumstances in which the effects of the excluded volume and of intramolecular interactions may be ignored in that the mean square end-to-end distance is equal to its unperturbed value.

6.6   Free spaces within a gel

A dilute gel contains a large amount of solvent so that the volume fraction of the polymeric material is small. An interesting problem concerns the fraction, f_r, of the total volume of the gel that is available for occupancy by a spherical particle of radius, r. It might be naively thought that this would be equal to $1 - \phi_2$ and this conclusion is certainly true for particles as small as the solvent molecules. For larger particles, however, the polymer chains of the gel are close together and so prevent the spherical particle from entering a certain volume situated between them.

The problem of calculating f_r is a problem of some difficulty, and has only been solved on the basis of a somewhat oversimplified model of the gel structure. In this model it is supposed that the polymer chains resemble rigid rods of length $2L$ distributed with their centres and orientations at random and with N_2 as the average number of centres per unit volume.

In terms of this model, Ogston has shown that

$$f_r = e^{-(2N_2 \pi L r^2 + 4\pi r^3/3)} \tag{6.25}$$

which for large values of L reduces to

$$f_r = e^{-2\pi N_2 L r^2} \tag{6.26}$$

It is interesting to note that this expression depends on $N_2 L$, the total length of all the fibres present, and not on their individual lengths. This should be proportional to ρ_2, the mass concentration of polymeric material in the gel.

If the polymer chains are stiff, it is reasonable to suppose that they behave as a collection of rod-like segments distributed at random with random orientations, provided the cross links do not disturb this random distribution. Thus the model used by Ogston may be expected to be valid for a gel in which the cross links have been formed between polymer molecules distributed at random. It is therefore gratifying that equation (6.26) does indeed describe the properties of a variety of particles in exclusion chromatography in a variety of gels such as polyacrylamide, agar or cross-linked polydextran.

A weakness of the theory is that it does not take explicit account of the number of cross links, and it is known from experiments using gels in exclusion chromatography that f_r also depends on the number of cross links introduced into the gel. In this context it is interesting to note that if the cross links are distributed at random in the volume occupied by the gel, and if the polymer chains connect up nearest neighbour cross links, there is a unique relation between the total concentration of cross links and the total concentration of polymer segments. If this relation is not satisfied and there are too many cross links, these must cluster together. There is indeed some evidence that in gels containing high concentrations of cross links the polymer chains are clustered together to form rope-like strands with relatively large voids between them. Thus f_r tends to be larger at high concentrations of cross links.

Problems

6.1   A rubber band maintained at 25° was extended to different lengths by the application of a force at one end, the other being fixed. The forces F, required to extend the band to a length, L, are tabulated below. Estimate the modulus of elasticity at small strain and at large strain. Is the ratio of these two moduli in accord with the value required by the simple theory of rubber elasticity discussed in the text?

$F(N)$	$L(cm)$
0.36	5.17
0.58	5.31
0.76	5.46
1.17	5.81
1.59	6.27
2.39	7.44
2.94	8.35
3.55	9.74
4.36	11.80
4.93	13.36

6.2   The radius of gyration and second virial coefficient of a sample of atactic polystyrene dissolved in cyclohexane were measured (by light-scattering methods) at a series of temperatures with the results tabulated below. Estimate the unperturbed root mean square end-to-end distance and the theta temperature in this solvent. What is the expansion factor, α, at 320 K?

$T(K)$	$s(nm)$	B(arbitrary units)
305.7	47.9	−0.40
307.2	51.8	−0.20
311.2	57.6	0.37
318.2	62.5	0.95
328.2	66.5	1.58

7 The scattering of radiation by macromolecules

BY S. D. DOVER

Electromagnetic radiation can interact with matter – macromolecules in particular – in either of two ways: by absorption or by scattering. In absorption, energy is taken from the incident beam. This energy may be dissipated so as to increase the thermal motion of the molecules in the solution; or it may be reemitted as fluorescence or phosphorescence at a longer wavelength at some later time; or it may suffer some alternative fate such as bringing about a photochemical reaction. In scattering, the radiation is merely deflected from its original path with little or no change in wavelength. In this chapter we shall be concerned solely with the scattering process.

We shall see that the characteristics of the scattered radiation depend on the size and shape of the scattering macromolecules and also on the details of its random Brownian motion. In this way we shall see that estimates of the molecular weight, radius of gyration and diffusion coefficient of a macromolecule can be obtained from measurements of this scattered radiation.

In order to understand the details of the scattering process we shall first review certain salient and relevant properties of electromagnetic radiation. We shall also mention and briefly discuss the scattering of neutrons by macromolecules.

7.1 Electromagnetic radiation

Electromagnetic radiation may be viewed as consisting of oscillatory electric and magnetic fields by which energy is transmitted through space. In the scattering process, only the electric field is of importance, so that the magnetic component may be ignored as far as the present discussion is concerned. For parallel plane-polarised and monochromatic radiation this electric field may be represented by

$$E = E_0 \cos 2\pi(\nu t - x/\lambda) \tag{7.1}$$

where we suppose that the radiation is travelling in the x direction. This relation shows that if we were to examine the electric field at a point fixed in space, the electric field strength would vary sinusoidally between $\pm E_0$ with

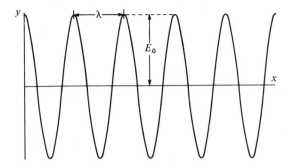

Figure 7.1. Graphical representation of a 'snapshot' of a plane-polarised monochromatic wave travelling in the x direction. The ordinate, E, represents the magnitude of the field strength at different points x. λ is the wavelength of the wave and E_0 is the amplitude. The diagram may also be taken to represent a graph of the field strength against time when observed at a fixed point. In this case the distance on the time scale between peaks is equal to $1/\nu$ where ν is the frequency.

time, with a frequency ν, as illustrated in figure 7.1. Alternatively if we were to obtain a 'snapshot' of the electric field at a certain instant of time it would be found to vary sinusoidally with x with a distance λ, the wavelength, between successive maxima. More generally the position in space at which the electric field was equal to some fixed value (E_0 for instance) moves with a velocity $c = \lambda\nu$, termed the phase velocity. This phase velocity depends on the refractive index of the medium through which the radiation is moving.

The electromagnetic spectrum spans an enormous range of wavelengths from small fractions of a nanometre for gamma rays to thousands of metres for radio waves. We shall consider only X-rays and visible radiation or light. A commonly used source of X-rays has a wavelength of 0.154 nm *in vacuo* and a commonly used light source produces a wavelength of 546 nm. Other regions of wavelength are not commonly used in scattering experiments, either because they are not appreciably scattered by macromolecules or because they are absorbed by virtually all solvents.

The parameter E_0 is termed the *amplitude* of the radiation and is a measure of the maximum value that the electric field adopts. E_0^2 is proportional to the rate of flow of the energy carried by the radiation (that is the amount of energy passing in unit time through a plane of unit area and perpendicular to x) or the *intensity* of the radiation.

For plane-polarised radiation the electric field is always perpendicular to x and in a fixed plane containing x, say the xy plane. We note in passing that, in circularly or elliptically polarised radiation, the electric field vector rotates about the x axis making complete revolutions with a frequency ν.

It is often convenient to rewrite equation (7.1) in the form

$$E = E_0 \cos{(\omega t - kx)} \tag{7.2}$$

where $\omega = 2\pi\nu$ is termed the angular frequency and $k = 2\pi/\lambda$. The quantity $\omega t - kx$ is now an angle. If the origin from which we measure t or x is changed we may allow for this by adding a further term to this angle called the phase angle or *phase*, ϕ. Thus, if the origin of x is moved a distance Δx, the change of phase angle is $k\Delta x$ or $2\pi\Delta x/\lambda$.

Two waves of the same wavelength may differ in phase. It is important for our present purposes to consider the resultant electric field due to the presence of two monochromatic waves both moving in the x direction and polarised with their electric vectors in the same plane but with a phase difference ϕ between them. Since the electric fields are parallel, their signed magnitudes can be added so that the resultant field is given by

$$E = E_1 \cos{(\omega t - kx)} + E_2 \cos{(\omega t - kx + \phi)} \tag{7.3}$$

If ϕ is an integral multiple of $360°$ the effect is as if one of the waves has been shifted along the x axis by an integral number of wavelengths with respect to the other so that the peaks and troughs of the two waves coincide. In this case the amplitude of the resultant is $E_1 + E_2$ and they are said to show *constructive interference*. If, on the other hand, ϕ is an odd multiple of $180°$, peaks coincide with troughs and the resultant amplitude is $E_1 - E_2$, which will be zero if $E_1 = E_2$. This corresponds to *destructive interference*. For intermediate phase differences, the amplitude depends on E_1, E_2 and ϕ.

The easiest way of calculating this resultant amplitude is by the use of complex numbers. We discuss this method more fully in appendix B. Here we note that the expression, $E_0 \cos{(\omega t - kx + \phi)}$ is equal to the real part of the complex number $E_0 \, e^{i(\omega t - kx + \phi)}$. Thus the resultant electric field given by equation (7.3) is equal to the real part of $e^{i(\omega t - kx)}(E_1 + E_2 \, e^{i\phi})$. If we suppose that this field may be represented by $E_r \, e^{i(\omega t - kx + \psi)}$ it is simple to determine the resultant amplitude, E_r, and phase, ψ, by equating the real and imaginary parts of the two expressions. This turns out to be much easier than a derivation conducted solely in terms of trigonometric functions. The resultant amplitude gives an intensity for the combination of the two waves equal to $E_1^2 + E_2^2 + 2E_1E_2 \cos{\phi}$. The term in $\cos{\phi}$ arises because the two original waves are *coherent;* that is to say there is a constant phase angle between them. This treatment can be extended to give the resultant of any number of coherent waves.

An interesting situation prevails if there are a large number of different waves with random phase angles; that is to say they are *incoherent*. In this case the average value of $\cos{\phi}$ is zero, and the resultant intensity is simply the sum of the intensities of the incoherent waves. This is the situation that prevails in ordinary light or X-ray beams. If, in addition, the directions of the planes of polarisation of the component waves are at random, the radiation is unpolarised.

7.2 The mechanism of scattering

We shall first consider the scattering of electromagnetic radiation by individual atoms. The source of the scattering is to be found in the electrons which surround the nucleus. The oscillating electric field, acting on the electrons, causes them to vibrate and to act as an oscillating dipole. Such an oscillating dipole reradiates the energy it has taken from the incident radiation. Since the electrons oscillate with the same frequency as the incident wave, their scattered radiation is also of the same frequency. Such scattering in which there is no change in the frequency is termed *elastic scattering*. The origin of this term derives from the alternative quantum mechanical point of view in which a quantum of light, acting like a particle, interacts 'elastically' with the electron with no loss of energy.

To investigate more closely the nature of the interaction between the oscillating field and the electron, we may suppose that the electron is bound to the nucleus by an elastic spring such that the restoring force is proportional to any small displacement that it may suffer from its equilibrium position. This is a valid approximation for the small displacements produced by light of ordinary intensity.

If somehow we were to displace such an electron and then release it, its equation of motion (in the absence of an oscillating field) would be

$$m\frac{d^2z}{dt^2} + fz = 0 \tag{7.4}$$

where m is its mass, f the force constant of the spring and z its displacement. This is the equation for simple harmonic motion, so that the particle would vibrate with a natural frequency, ω_0, equal to $(f/m)^{1/2}$. If we now apply an oscillating electric field (implied by the arrival of the electromagnetic wave), $E_0 \cos \omega t$, the equation of motion becomes

$$m\frac{d^2z}{dt^2} + m\omega_0^2 z = qE_0 \cos \omega t \tag{7.5}$$

In practice an additional term must be incorporated in this equation to take account of the damping of the oscillations of the electron. With the damping term included, equation (7.5) would imply that the electron will start to vibrate with its natural frequency ω_0 when the wave first arrives. These oscillations will quickly die away and thereafter the electron will vibrate with the frequency of the exciting wave and its displacement will be given by

$$z = \frac{qE_0}{(\omega^2 - \omega_0^2)\,m} \cos \omega t \tag{7.6}$$

Two situations are of importance:

(i) $\omega \ll \omega_0$ so that $z = -\dfrac{qE_0}{\omega_0^2 m} \cos \omega t$ (7.7)

(ii) $\omega \gg \omega_0$ so that $z = \dfrac{qE_0}{\omega^2 m} \cos \omega t$ (7.8)

We may note in passing that if ω is close to ω_0 absorption takes place, the model of an electron bound by a spring becomes inapplicable and quantum mechanical considerations must be invoked. We shall not consider such possibilities.

The natural frequency of the electrons in many atoms and molecules (which is the frequency of the absorption bands) lies in a region somewhere between that of X-rays and light. Thus light scattering corresponds to the first situation and X-ray scattering corresponds to the second situation. This constitutes one major difference between light and X-ray scattering.

Now the instantaneous magnitude of the dipole associated with the displaced electron is $p = qz$, and it may be shown that the amplitude of the radiation emitted by an oscillating dipole is proportional to $d^2 p / dt^2$. We shall see later that it also depends on the direction. We thus see that the field of the scattered radiation is proportional in our cases of interest to

(i) $\dfrac{d^2 p}{dt^2} = \dfrac{-qE_0 \omega^2}{\omega_0^2 m} \cos \omega t$ (7.9)

(ii) $\dfrac{d^2 p}{dt^2} = \dfrac{qE_0}{m} \cos \omega t$ (7.10)

Since the intensity of the scattered radiation is proportional to the square of its amplitude, we see that in the case of light scattering the intensity is proportional to ω^4 or λ^{-4}, whereas in the case of X-ray scattering it is independent of the frequency or wavelength of the incident radiation. This constitutes the second major difference between the two sorts of scattering.

A third major difference is provided by the fact that the wavelength of X-rays is small compared to the size of the electron cloud of an atom. This means that a particular peak or trough of the incident wave will reach different parts of the electron cloud at different times. This implies that the oscillating dipoles set up at different parts of the cloud will not be in phase and hence that the different waves scattered from different parts of the cloud will have phase differences between them, although, in so far as they are all due to the same incident wave, they will be coherent. It may also be seen, as illustrated in figure 7.2, that the waves scattered from different parts, and which reach a certain point of observation, will have travelled different distances and hence suffered a further phase change. This latter effect depends on the angle at which we observe them. The amplitude of the wave we observe

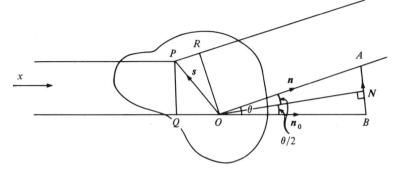

Figure 7.2. The incident wave travelling in the x direction impinges on two elements of a macromolecule at O and at P. The vector s is the displacement of P from O. n_0 and n are two unit vectors in the directions of the incident wave and a scattered wave at scattering angle θ. The lengths of OA and OB are both unity. N is the vector difference $n - n_0$. Its length is seen from the triangle OAB to be equal to $2 \sin \frac{1}{2} \theta$.

will then depend on the interference between all these waves and will depend on the angle. The net result of these interference effects is that the intensity falls off, even for scattering from a single atom, with the scattering angle as depicted in figure 7.3.

The situation is different in light scattering. In this case the wavelength is large compared with the size of the atom and all the dipoles and the scattered wavelets are essentially all in phase. This is true even for the smaller macromolecules and it is only for very large macromolecules, the size of whose domain is comparable to or larger than the wavelength of the light (about 500 nm), that such interference effects become important.

The scattering of neutrons is somewhat different. Although these are usually thought of as particles, they also behave as waves. In this case their wavelength is determined by their momentum, which depends on their velocity. For a neutron wave there is nothing that corresponds to the electric field manifesting a direction so that they do not exhibit polarisation (this is not strictly true for, in a magnetic field, their spins may be 'polarised', but this is not relevant in the present context). Neutrons do not interact appreciably with electrons, so that their scattering is the result of their interaction with the nuclei of the atoms they encounter. Since the wavelength of the neutrons usually encountered in scattering experiments is large compared to the size of an atomic nucleus, the interference effects encountered in X-ray scattering are not encountered in the scattering of neutrons from a single atom, and the scattering intensity does not fall off with angle as in figure 7.3. The details of the scattering process are not fully understood, and it must be accepted as

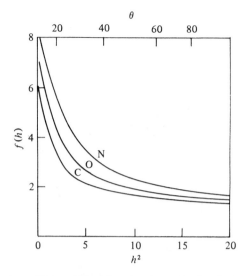

Figure 7.3. The atomic scattering factors, $f(h)$, for carbon, nitrogen and oxygen for $\lambda = 0.154$ nm expressed in terms of the amplitude scattered by a free electron. The intensity falls off as $\{f(h)\}^2$, and h^2 is equal to $16\pi^2 \sin^2 \frac{1}{2}\theta / \lambda^2$, where θ is the scattering angle.

an empirical fact that the scattering power of most atoms (C,O,N for instance), including deuterium, are similar in magnitude, whereas that of hydrogen is somewhat different. Finally we should note that there is no analogue of the induced dipole, so that the angular dependence of the dipole radiation discussed in the next section is absent, and neutrons are scattered equally in all directions by a single atom.

7.3   The angular distribution of intensity of dipole radiation

The interference effects mentioned in the previous section are due to the interaction of radiation scattered by dipoles in different parts of the electron cloud or molecule; but these dipoles themselves radiate with an angular variation.

Let us consider the amplitude of the radiation due to a single dipole in a direction defined by the angles ϕ and θ in figure 7.4. The direction of the dipole is an axis of symmetry of the distribution of radiation, so that the amplitude is independent of θ. Nevertheless the amplitude does depend on ϕ. Electromagnetic theory tells us that the amplitude is zero in the direction of the dipole for $\phi = 90°$ and maximal in a direction perpendicular to it for $\phi = 0$. To obtain the amplitude at intermediate angles, we may observe that the dipole (which as indicated in chapter 2 is a vector) can be resolved into

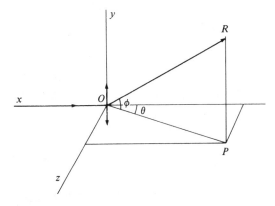

Figure 7.4. A wave, polarised in the xy plane, travels in the x direction and induces a dipole orientated in the y direction at O. A wave is scattered in the direction OR. The point P is in the xz plane.

two components: a component $p \cos \phi$ perpendicular to the direction defined by ϕ and a component $p \sin \phi$ in a parallel direction. The amplitude due to the latter component is zero, so that the only contribution to the total is due to the former component. Thus the amplitude in the directions defined by ϕ is proportional to $\cos \phi$.

If the incident radiation is plane polarised, all the dipoles lie in the plane of polarisation and perpendicular to x so that the intensity scattered at an angle ϕ is proportional to $\cos^2 \phi$. It is, however, more convenient and usual from an experimental point of view to use unpolarised incident radiation. Unpolarised radiation consists of a large number of wave components making different angles, ψ, with the y direction with no preferred angle ψ. It therefore induces a large number of dipoles orientated at different angles ψ. The proportion of the total energy of the beam, and hence the proportion of the total number of dipoles that are orientated in the range ψ to $\psi + d\psi$, is independent of ψ and equal to $d\psi/2\pi$. Furthermore the different wave components are mutually incoherent; this means that we should add the intensities of radiation scattered in a particular direction and due to the different dipoles to get the total intensity in that direction.

We may now observe that, for unpolarised radiation, since no direction ψ is preferred, the direction, x, of the incident wave is an axis of symmetry. This means that the intensity at any point on the circle illustrated in figure 7.5 is the same and hence that it is independent of the angle ψ, and the intensity in the direction OP' will be the same as in the direction OP as defined in figure 7.5.

To find the intensity along OP' let us consider a single dipole orientated at an angle ψ. This may be resolved into two components: $p \cos \psi$ in the y

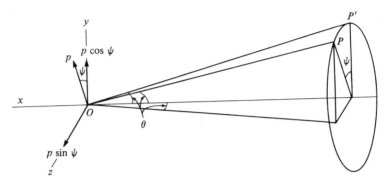

Figure 7.5. For unpolarised incident radiation travelling in the x direction, the latter is an axis of symmetry. Thus the scattered intensity is the same at every point on the circle defined by the angle θ and is independent of ψ. A dipole induced at O at an arbitrary angle ψ to the y axis may be resolved into two components, $p \cos \psi$ and $p \sin \psi$, in the y and z directions. The former component makes an angle $90° - \theta$ to OP' and the latter is perpendicular to OP'.

direction and $p \sin \psi$ in the z direction. The intensity scattered by the former component along OP' is proportional to $\cos^2 \psi \cos^2 \theta$ and that due to the latter is proportional to $\sin^2 \psi$ (this component is perpendicular to OP'). Thus the total intensity along OP' due to both components is proportional to $\cos^2 \psi \cos^2 \theta + \sin^2 \psi$. We then see that the total intensity along OP' due to all the dipoles distributed over different angles ψ is proportional to

$$\frac{1}{2\pi} \int_0^{2\pi} (\cos^2 \psi \cos^2 \theta + \sin^2 \psi) \, d\psi$$

This expression is equal to $\frac{1}{2}(1 + \cos^2 \theta)$.

To complete the geometrical picture we now note that the waves scattered in any direction are spherical so that the amplitude decreases with distance, d, from the radiating dipole. Thus the intensity of the scattered radiation at an angle θ and at a distance, d, is proportional to $\frac{1}{2}(1 + \cos^2 \theta)/d^2$ for unpolarised incident radiation. The constant of proportionality is, we must remember, in turn proportional to an angle-dependent atomic scattering factor to take into account interference effects discussed in the previous section. For scattering from particles that are small compared to the wavelength, this scattering factor is unity.

7.4   The amplitude and intensity of scattered light

We will now consider specifically the scattering of light for which $\omega \ll \omega_0$ and not return to X-ray scattering until section 7.9.

In the last section we have seen that the amplitude of the light scattered at an angle ϕ from a plane-polarised incident beam and observed at a distance, d, from the scattering dipole is proportional to $\cos \phi/d$. We have also seen in section 7.2 that the electric field of the radiation emitted by an oscillating dipole is proportional to d^2p/dt^2. We may thus see that the field associated with the light scattered by a *point scatterer*, which is small compared to the wavelength, is given by

$$E_s = \frac{1}{\epsilon_0 c^2} \frac{d^2p}{dt^2} \frac{\cos \phi}{d} \qquad (7.11)$$

where the constant of proportionality given by theory is $1/\epsilon_0 c^2$; c is the velocity of light and ϵ_0 the permittivity of free space.

If we were concerned with a single electron we could substitute for d^2p/dt^2 from equation (7.9), but in reality we are interested in scattering by a whole atom or molecule so that we will proceed in a different manner. In section 2.3 we have defined the polarisability, α, of a molecule and may thus see that the dipole induced by the oscillating field of the incident wave is given by $\alpha E_0 \cos (\omega t - kx)$. If we differentiate this twice with respect to t and substitute into equation (7.11) we obtain

$$E_s = \frac{4\pi^2 \alpha E_0}{\epsilon_0 \lambda^2 d} \cos \phi \cos (\omega t - kx) \qquad (7.12)$$

where we have replaced ω by $2\pi c/\lambda$. This result shows that the scattered wave has the same frequency as the incident light. The intensity of this wave is, as usual, given by the square of the amplitude.

So much for the intensity scattered by a single molecule. Let us now suppose that there are N_s point scatterers situated at random (such as might be the case in a perfect gas) and moving at random. The waves scattered from these will have no consistent phase relationship and thus be mutually incoherent. Then the total intensity scattered by the N_s molecules will be N_s times that scattered by a single one. If the weight concentration is ρ, there will be $N_A \rho/M$ molecules per unit volume so that the intensity scattered by unit volume will be $N_A \rho/M$ times that scattered by a single molecule. M is the molar mass.

If instead of considering the scattering produced by polarised incident radiation we consider that produced by unpolarised light, the factor $\cos^2 \phi$ in the expression for the intensity must be replaced by $\frac{1}{2}(1 + \cos^2 \theta)$. We thus obtain the final result

$$I_s/I_0 = \frac{8\pi^4 \alpha^2 (1 + \cos^2 \theta) N_A \rho}{\epsilon_0^2 \lambda^4 d^2 M} \qquad (7.13)$$

where I_0 is the intensity of the unpolarised incident light and I_s is the intensity of light scattered from unit volume and observed at an angle θ to the direction of the incident light and at a distance, d.

Scattering of this nature from a collection of random point scatterers is termed *Rayleigh scattering*. If is convenient to lump the intensities and geometry-dependent terms into a single expression, the *Rayleigh ratio*, R_θ, so that

$$R_\theta = \frac{I_s d^2}{I_0(1 + \cos^2 \theta)} = \frac{8\pi^4 \alpha^2 N_A \rho}{\lambda^4 M \epsilon_0^2} \tag{7.14}$$

R_θ should then be independent of θ for point scatterers. For large molecules, such that interference effects are relevant, R_θ will be found to vary with θ.

7.5   Scattering from condensed systems

Equation (7.14) applies to scattering from a dilute gas composed of point scatterers in which the waves scattered by the different molecules are mutually incoherent. If we now consider an opposite extreme in which the molecules are regularly arranged in a crystal lattice, there will be no scattering at all (except from its surfaces). This is because, for any molecule in the lattice, there will be another one fixed relative to the first such that there will be a constant phase difference of $180°$ between the two waves scattered by the pair at any angle except $\theta = 0$. These two waves will interfere to give complete destructive interference.

A perfectly homogeneous solvent would behave similarly for we may divide it up into a large number of identical volume elements, and the waves scattered from these suffer destructive interference in pairs.

Light is only scattered from real solvents and solutions because they are not perfectly homogeneous; that is to say the instantaneous polarisability of any one of the volume elements fluctuates randomly with time. Let us suppose that the instantaneous value of the polarisability of one of the elements is $\alpha = \alpha_0 + \delta\alpha$, where α_0 is the average polarisability which is the same for all the elements. The intensity of light scattered at a given time by one of the elements will then be proportional to α^2 as implied by equation (7.12). Since α^2 may be written as $\alpha_0^2 + 2\alpha_0 \delta\alpha + (\delta\alpha)^2$ the light scattered by each element is the sum of three terms. The first term, proportional to α_0^2, is the same for all elements and for this reason the corresponding components of pairs of elements disappear by destructive interference as for a homogeneous medium. This leaves the other two components to be considered, which arise from the fluctuations $\delta\alpha$. The average intensity scattered by an element is then seen to be proportional to the average $\langle 2\alpha_0 \delta\alpha + (\delta\alpha)^2 \rangle$. Since positive and negative values of the fluctuation $\delta\alpha$ are equally unlikely, $\langle \delta\alpha \rangle = 0$. We are thus left with only the third component as contributing to the average intensity scattered by a single element. Since these fluctuations are random there is no constant phase relation between these remaining components of the waves scattered by the different elements. The intensities of these components can therefore be summed over all the volume elements to obtain the total scattered intensity.

From a thermodynamic point of view, the properties of a two-component solution are fixed once the pressure, temperature and solute concentration have been fixed. This implies that any fluctuation, $\delta\alpha$, in the polarisability may be viewed as being caused by fluctuations δT in the temperature, δP in the pressure or $\delta\rho_2$ in the concentration of the solute. Thus we may write

$$\delta\alpha = \left(\frac{\partial\alpha}{\partial P}\right)_{T,\rho_2} \delta P + \left(\frac{\partial\alpha}{\partial T}\right)_{P,\rho_2} \delta T + \left(\frac{\partial\alpha}{\partial\rho_2}\right)_{P,T} \delta\rho_2 \qquad (7.15)$$

The fluctuations, δP and δT, in the pressure and temperature will be closely similar in the solution and in the pure solvent so that the scattering intensity arising in the pure solvent and in the solution from these fluctuations will be essentially the same. If we restrict our attention to the difference in the intensities scattered by the solvent and solution we need only consider fluctuations in the concentration. This difference in intensities is termed the *excess scattering intensity*. We thus see that we may write

$$\langle(\delta\alpha)^2\rangle = \left(\frac{\partial\alpha}{\partial\rho_2}\right)_{T,\rho_2}^2 \langle(\delta\rho_2)^2\rangle \qquad (7.16)$$

Now the polarisability, α, of a volume element, v, is related to the refractive index of the solution, n, by

$$n^2 - 1 = 4\pi\alpha/v\epsilon_0 \qquad (7.17)$$

If we differentiate this expression with respect to the concentration of the solute we obtain

$$\frac{\mathrm{d}\alpha}{\mathrm{d}\rho_2} = \frac{v\epsilon_0 n_0}{2\pi} \frac{\mathrm{d}n}{\mathrm{d}\rho_2} \qquad (7.18)$$

where we have replaced n by n_0, the refractive index of the solvent, for in dilute solutions the two are almost the same. $\mathrm{d}n/\mathrm{d}\rho_2$ is the *refractive index increment* of the solute (in the particular solvent of interest) and is readily measurable. It should be noted that the refractive index, n_0, and the refractive index increment vary with the frequency of the light. The correct values to employ are those that correspond to the frequency of the incident light, $\nu\ (=\omega/2\pi)$.

We have already considered in chapter 3 the nature of concentration fluctuations and equation (3.56) gives an expression for the mean square concentration fluctuation in a small volume element v.

We are now in a position to write an expression for the Rayleigh ratio for light scattered at an angle θ from unit volume of a solution of concentration ρ_2 and observed at a distance, d. We must bear in mind that the intensity, I_s, in the definition of the Rayleigh ratio (equation (7.14)) is the excess intensity. First we replace $N_A \rho/M$ in equation (7.14), which is the number of scattering elements in a unit volume, by $1/v$; next we replace α^2 by $\langle(\delta\alpha)^2\rangle$. We then substitute from equation (7.16) for $\langle(\delta\alpha)^2\rangle$, from (7.18) for $d\alpha/d\rho_2$ and from equation (3.56) for $\langle(\delta\rho)^2\rangle$ to obtain the final expression

$$R_\theta = \frac{2\pi^2 n_0^2 (dn/d\rho_2)^2 \, \rho_2}{\lambda^4 N_A (1/M_2 + 2B\rho_2 + \cdots)} \tag{7.19}$$

7.6   The measurement of molecular weights by light scattering

Equation (7.19) may be written in the form

$$\frac{K\rho_2}{R_\theta} = \frac{1}{M_2} + 2B\rho_2 + \cdots \tag{7.20}$$

where K is a constant containing all the 'optical' parameters and is equal to $2\pi^2 n_0^2 (dn/d\rho_2)^2/N_A\lambda^4$. Since all the parameters on the left-hand side of equation (7.20) are open to experimental measurement, they may be used to obtain estimates of the molecular weight of the solute.

The usual procedure is to measure the scattering ratio I_s/I_0 at some convenient angle θ (usually $90°$ so that $\frac{1}{2}(1 + \cos^2\theta) = 1$) for a series of solutions of different concentration, ρ_2, and to plot $K\rho_2/R_\theta$ against ρ_2. The molar mass is then given by the intercept at $\rho_2 = 0$. The second virial coefficient B may also be obtained from the slope of such a plot. It should be remembered that equation (7.20) only applies to macromolecules the size of whose domain is small compared to the wavelength of the light that is used.

It is important to note that the scattered intensity, I_s, that is used in the calculation of R_θ is the intensity scattered from unit volume of the solution. In practice an arbitrary volume, V, is illuminated by the incident light so that the measured intensity must be divided by V to obtain I_s. V is often measured by determining the intensity scattered by a solution whose scattering ratio has been carefully measured previously, or is otherwise known. In this way the apparatus can be calibrated.

One point has been glossed over in the derivation of equation (7.20). That is, we have assumed that the scattering element is isotropic, which is to say that its polarisability is the same in all directions. If this is not so, the dipole induced by the incident light may not be parallel to the incident electric field. If we observe the light scattered horizontally (z direction in figure 7.5) at $\theta = 90°$ from unpolarised incident light, it will normally be polarised in the vertical plane. With anisotropic scatterers there will be a small component which is horizontally polarised. The ratio of the horizontally polarised component to the vertically polarised component defines the *depolarisation ratio*, ρ_d. Cabannes showed that the Rayleigh ratio, R_θ, must be multiplied by $(6 - 7\rho_d)/(6 + 6\rho_d)$ to allow for this effect.

If the macromolecule is polydisperse, the measured scattered intensity will be the sum of the intensities scattered by all the macromolecular species present. If the optical constant K is the same for all macromolecules, and $\rho = \Sigma_i\,\rho_i$ is written for the total concentration of all species, we see from equation (7.20) that the limit of $K\rho/R_\theta$ as measured is equal to $\Sigma_i\,\rho_i/\Sigma_i\,M_i\rho_i$ which is equal to $1/M_w$, where M_w is the weight-average molar mass.

Thus similar information may be obtained from measurements of the osmotic pressure and of light scattering but, whereas the former gives the number-average molecular weight, the latter gives the weight average.

It is important to note that the intensity of the scattered light is inversely proportional to the molecular weight. This means that the scattered intensity may be too small to measure with adequate precision for small macromolecules. It also means that large errors may be introduced if the solution contains large dust particles and rigorous steps must be taken to remove these. With the advent of the laser light sources, very small scattering volumes may be used. If there are relatively few dust particles present, these will float in and out of the light beam and it is sometimes possible to recognise their presence and to discard intensity measurements that are anomalously high because of it.

7.7 Light scattering from larger particles

We must now consider the scattering of light from larger particles. As we have discussed in section 7.2, the oscillating dipoles at two points in a macromolecule, O and P (see figure 7.2) will not, in general, be in phase since the peak of a wave will reach the two points at different times. This phase difference, measured as an angle, is given by $2\pi/\lambda$ times the distance OQ. Similarly the light scattered at O and P at an angle θ will reach the observer at different times so that a further phase difference is introduced that depends on θ and which is represented by $2\pi/\lambda$ times PR. The two waves scattered at O and P will be coherent so that their amplitudes must be added taking into account the phase difference as discussed in section 7.1. We must in fact add the contributions scattered from all parts of the macromolecule to obtain the total amplitude experienced by the observer.

To do this it is convenient to define two unit vectors, the first n_0 in the direction of the incident light and the second n in the direction θ. By unit vectors we mean vectors whose lengths, written for instance as $|n|$, are equal to unity. We also define a vector s to describe the position of a point on a macromolecule relative to an origin O. With reference to figure 7.2 we then see that the distance OQ may be written $n_0 \cdot s$ and PR as $n \cdot s$. Thus the phase difference of the light scattered from P relative to that scattered from the origin is $2\pi(n - n_0) \cdot s/\lambda$, or $2\pi N \cdot s/\lambda$ where N is the vector difference $n - n_0$. It is easy to see from the triangle of vectors in figure 7.2 that the length of N is $2 \sin \frac{1}{2}\theta$.

Let us now suppose that the amplitude of the wave scattered at P is a. We may then represent the electric field due to this wave by the real part of $a\, e^{i(\omega t-kx+2\pi N\cdot s/\lambda)}$. If we suppose that the macromolecule is composed of a number of elements located at points represented by s_i and scattering wavelets of amplitude a_i, the total field at an angle θ resulting from all these wavelets is given by the real part of $\Sigma_i\, a_i\, e^{i(\omega t-kx+2\pi N\cdot s_i/\lambda)}$. We may note that $e^{i(\omega t-kx)}$ is a constant factor in all the terms in this sum; it represents the oscillating part of the expression for the field so that the amplitude of the resultant field is represented simply by $\Sigma_i\, a_i\, e^{2\pi i N\cdot s_i/\lambda}$. It is convenient to define a vector $h = 2\pi N/\lambda$ so that the resultant amplitude may be written $\Sigma_i\, a_i\, e^{ih\cdot s_i}$.

As indicated in appendix B, the intensity of this resultant is given by $\Sigma_i\, a_i\, e^{ih\cdot s_i}\, \Sigma_j\, a_j\, e^{-ih\cdot s_j}$ and this may be written as $\Sigma_i\, \Sigma_j\, a_i a_j\, e^{ih\cdot(s_i-s_j)}$. In this double summation, for every term in $s_k - s_l$, there will be a term in $s_l - s_k$ so that the resultant intensity may be written $\frac{1}{2} \Sigma_i\, \Sigma_j\, a_i a_j [e^{ih\cdot(s_i-s_j)} + e^{-ih\cdot(s_i-s_j)}]$ which is equal to $\Sigma_i\, \Sigma_j\, a_i a_j \cos[h\cdot(s_i-s_j)]$, which is a real quantity.

This intensity is that which would be observed if the macromolecule were to be fixed in space. In reality it is rotating in a random manner owing to its Brownian motion. It is reasonable to assume that the wave scattered when it is in one orientation will have no constant phase relation to that scattered when it is in another, so that we may take the average of the intensities for all possible orientations to obtain the intensity that we would observe. To do this we may average each of the terms $\cos[h\cdot(s_i-s_j)]$ over all orientations and sum the results. Now the vector h is fixed in space and the vector $s_i - s_j$ defines a direction fixed to the macromolecule. Let us suppose that the angle between h and $s_i - s_j$ is ϕ; then $h\cdot(s_i-s_j)$ is equal to $|h|\,|(s_i-s_j)|\cos\phi$.

Since all orientations of the macromolecule are equally likely, the probability that the angle between h and $s_i - s_j$ lies in the range ϕ to $\phi + d\phi$ is proportional to $\sin\phi\, d\phi$ as illustrated in figure 7.6. We thus see that our average intensity is given by

$$
\left\langle \sum_i \sum_j a_i a_j \cos[h\cdot(s_i-s_j)] \right\rangle
$$

$$
= \frac{\displaystyle\sum_i \sum_j a_i a_j \int_0^\pi \cos[|h|\,|s_i-s_j|\cos\phi]\, \sin\phi\, d\phi}{\displaystyle\int_0^\pi \sin\phi\, d\phi} \tag{7.21}
$$

This expression evaluates to $\Sigma_i\, \Sigma_j\, a_i a_j \sin[|h|\,|(s_i-s_j)|]/|h|\,|(s_i-s_j)|$, which may be expressed as a power series in the length of the vector, h, or h which is equal to $(4\pi \sin\frac{1}{2}\theta)/\lambda$. Taking the first two terms we find that the average amplitude is equal to $\Sigma_i\, \Sigma_j\, a_i a_j(1 - h^2 r_{ij}^2/6)$ where $r_{ij} = |s_i - s_j|$.

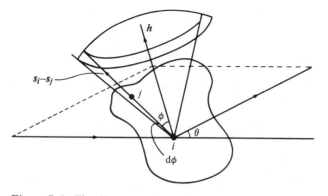

Figure 7.6. The direction of the vector h is defined by the angle of scatter, θ. The vector difference $s_i - s_j$ represents the displacement of two elements i and j of the macromolecule. Its direction changes as the molecule rotates. ϕ is the angle between h and $s_i - s_j$. The latter vector can point to any point on the sphere drawn with centre at i with equal probability. Thus the probability of it lying within unit solid angle is $1/4\pi$. Thus the probability of it making an angle in the range ϕ to $\phi + d\phi$ with h is proportional to the area of the annular surface element of the sphere as indicated in the diagram. The area of this surface is proportional to $\sin \phi \, d\phi$.

If the distances r_{ij} are all small compared with the wavelength, λ, the expression reduces to $\sum_i \sum_j a_i a_j$; at the same time the molecule becomes a point scatterer to which the expressions derived in section 7.4 for Rayleigh scattering apply. We thus see that the correction factor which must be applied to the scattered intensity to account for the interference effects with large particles is given, for small scattering angles, by

$$P(\theta) = 1 - \frac{\sum_i \sum_j a_i a_j h^2 r_{ij}^2 / 6}{\sum_i \sum_j a_i a_j} \tag{7.22}$$

If we take as origin (the point O in figure 7.4) the 'centre of gravity' of the scattering amplitudes of the elements composing the macromolecule so that $\sum_i a_i s_i = 0$, we find that the ratio of sums on the right-hand side of equation (7.22) is equal to $2h^2 \sum_i a_i s_i^2 / 6 \sum_i a_i$, where s_i is the distance from the 'centre of gravity' to the element i. If the macromolecule is a homopolymer for which each monomer unit is small compared to λ, and all the a_i are identical, this expression is identical to the square of the radius of gyration, as defined in section 4.2, times $h^2/3$. For a flexible polymer the mean square radius of gyration, $\langle s^2 \rangle$, should be used. For a heteropolymer, in which the monomer units are not identical, each distance s_i^2 must be weighted by the value of a_i appropriate to the unit, and the expression will not be equal to the true geometric squared radius of gyration. In practice errors introduced by ignoring this subtlety are small.

So far we have not considered the physical meaning of the amplitudes, a_i, but it is clear from section 7.4 that they are proportional to the polarisability, α_i, of the element of the isolated macromolecule.

For condensed systems, however, the situation is a little different. Here the scattering unit, as we discussed in section 7.5, is a small volume element of the solution. It is this that must be divided into a large number of smaller elements which scatter with amplitudes a_i. Some of these smaller elements will coincide with portions of the solution and some to elements of a macromolecule that is present in the volume element as a result of a concentration fluctuation. For reasons explained in section 7.5, the scattering due to the uniform background solution may be ignored so that only the contribution from the latter need be considered. If we now consider the excess scattered intensity only, we see that the α_i for each small element that contributes is proportional to $\alpha_i - \alpha_{i,0}$ where α_i is the polarisability of the element of the macromolecule and $\alpha_{i,0}$ is the polarisability of the element of solvent that the macromolecule has displaced. Again, for a homopolymer $\alpha_i - \alpha_{i,0}$ will be the same for each monomer unit, but in a rigorous treatment this must be taken into account in defining a radius of gyration.

An important point emerges if $\alpha_i = \alpha_{i,0}$ for all elements of the macromolecule. This corresponds to the circumstance that the polarisability of a volume element is independent of the number of macromolecules it contains so that $d\alpha/d\rho_2 = dn/d\rho_2 = 0$. Under these circumstances, as equation (7.19) above shows, there is no scattering to consider. Another way of saying this is that there is no *contrast* between solvent and solute.

Putting these observations together we then readily see for condensed systems that the intensity I_s of the light scattered in the direction θ, must be multiplied by $P(\theta)$ as given by equation (7.22), and that equation (7.20) must be written, for large particles, in the form

$$\frac{K\rho_2}{R_\theta} = \left[\frac{1}{M_2} + 2B\rho_2 + \cdots \right] \left[1 + \frac{16\pi^2 \langle s^2 \rangle \sin^2 \frac{1}{2}\theta}{3\lambda^2} \right] \tag{7.23}$$

for a flexible polymer (note that we have used the approximation $(1 - x)^{-1} \approx (1 + x)$ in obtaining equation (7.23)), while, for rigid macromolecules, $\langle s^2 \rangle$ must be replaced by the square of their uniquely defined radius of gyration.

7.8   The measurement of the radius of gyration

The radius of gyration of a macromolecule can be estimated from light-scattering measurements with the aid of equation (7.23). It is necessary to measure the excess scattered intensity, in order to calculate R_θ, for a series of different solutions of different concentrations. For each solution it is

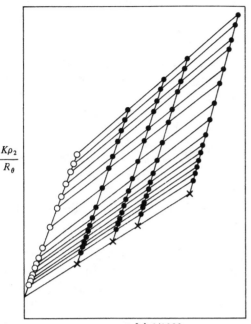

$$\frac{K\rho_2}{R_\theta}$$

$\rho_2 + \sin^2 \frac{1}{2}\, \theta/1000$

Figure 7.7   Zimm plot for cellulose nitrate in acetone. The solid circles correspond to experimental values of $K\rho_2/R_\theta$ plotted against $\rho_2 + \sin^2 \frac{1}{2}\theta/1000$. Four different concentrations were used and R_θ was measured for each at 15 different scattering angles θ. The open circles were obtained by extrapolating the experimental points corresponding to each angle to zero concentration. They lie on a line of slope equal to $16\pi^2 \langle s^2 \rangle/3000\lambda^2 M$, and meet the ordinate axis at an intercept equal to $1/M$. The crosses were obtained by extrapolating the experimental points corresponding to each angle to zero $\sin^2 \frac{1}{2}\theta$. They lie on a line with slope equal to $2B$ and again intercept the ordinate at a value of $1/M$. The diagram is taken from G. Oster in *Physical Methods of Chemistry*, part III-A (edited by Weissberger and Rossiter, Wiley, N.Y., 1972).

necessary to obtain the value of R_θ at several different scattering angles θ. It is then necessary to extrapolate $K\rho/R_\theta$ to zero concentration for each angle. B and M may also be found by extrapolating to zero angle for each concentration.

This double extrapolation is conveniently performed with the aid of a *Zimm plot*. In this a parameter, X, is calculated for each measurement from the relation $X = \rho_2 + k \sin^2 \frac{1}{2}\theta$, where k is some constant which may be chosen quite arbitrarily in order to make the plot reasonable in scale. $K\rho_2/R_\theta$ is then plotted against X as illustrated in figure 7.7 to form a grid of points.

Points corresponding to different values of ρ_2, the concentration, but identical values of θ, the scattering angle, are then used to define a line, in principle linear at low ρ_2, which is extrapolated to yield a value of $K\rho_2/R_\theta$ at zero concentration but corresponding to the same value of θ. This is done for each value of θ. These extrapolated points should then lie on a line whose slope (at low values of $\sin^2 \frac{1}{2}\theta$) is equal to $16\pi^2 k\langle s^2\rangle/3\lambda^2 M_2$ from which the radius of gyration can be estimated given M_2, the molar mass. The intercept of the line at $X = 0$ gives an estimate of $1/M_2$. The reverse procedure can then be followed to obtain an estimate of B, the second virial coefficient. To do this, points of constant ρ_2, but different $\sin^2 \frac{1}{2}\theta$, define lines which are extrapolated to zero $\sin^2 \frac{1}{2}\theta$. These extrapolated points should then lie on a straight line of slope $2B$ and again should extrapolate at zero X to give an estimate of $1/M_2$.

The success of this whole operation requires in the first place that R_θ varies significantly with θ. We have seen that this is not so for point scatterers for which R_θ is independent of θ. In practice this means that the radius of gyration must be greater than $\lambda/20$, and the method is inapplicable to macromolecules that are too small. It is also required that the approximation involved in equation (7.22) should be valid so that $K\rho_2/R_\theta$ is linear in $\sin^2 \frac{1}{2}\theta$. For very large particles the approximation breaks down, terms in higher powers of h must be incorporated into equation (7.22) and the extrapolation to zero $\sin^2 \theta$ becomes too uncertain except at very low scattering angles. This puts a practical upper limit on the radius of gyration of $\lambda/2$.

Finally it should be noted that the z-average radius of gyration is obtained if the method is applied to polydisperse preparations.

7.9   Low angle X-ray scattering

If we return to equations (7.9) and (7.10) we may note in the case of light scattering (equation (7.9)) that the amplitude of the oscillating dipole depends on ω_0^2 which, in turn, depends on the force constant of the 'spring' which attaches the electron to the molecule. Such molecular details are not readily open to experimental measurement and it was for this reason that we simplified matters by invoking the polarisability α. With X-rays, the situation is simpler and we see that the amplitude of the oscillating dipole of a single electron is equal to qE_0/m irrespective of how 'tightly' it may be bound to the atom. Thus the scattering power of an isolated atom is proportional to the number of electrons it contains. In condensed systems it is proportional to the difference in the number of electrons it carries and the number of electrons carried by the solvent displaced. Thus the contrast depends on the difference in electron density of the solvent and solute. We may of course, if we wish, define the X-ray polarisability of an atom as qZ/m and this in turn may be related to the refractive index. If we do this an interesting point emerges by comparing equations (7.9) and (7.10), when we observe that the sign of the two expressions for d^2p/dt^2 are different. This implies that the polarisability for X-rays is negative and hence (from equation (7.17)) that the refractive index is less than unity. This implies that the phase velocity of X-rays in a medium is greater than its value in free space!

The next important point is that the wavelength of X-rays is less than the diameter of the atom. This means that the wavelets scattered from different parts of the atom interfere in a manner described approximately by equation (7.22) at low scattering angles. Figure 7.3 shows how the atomic scattering factor, $f(h)$, varies with $h^2 (= 16\pi^2 \sin^2 \frac{1}{2}\theta/\lambda^2)$ for several common atoms. Since the wavelength of X-rays is much smaller than that of light, the value of h^2 for a given scattering angle, θ, is much greater: in fact, by a factor of the order of 10^7. This has important implications when we consider the X-ray scattering of molecules.

If we take into account the modifications pertinent to X-rays that we have noted above, the theory of X-ray scattering proceeds along lines analogous to light scattering. The scattering amplitudes, a_i, in section 7.7 must be replaced by the atomic scattering factors, $f(h)$, and it may be noted, since only the square of the polarisability enters into the results, these are not affected by the fact that α is negative. The radius of gyration must be defined in terms of the excess electron density (rather than polarisability) though for homopolymers this does not alter the final expression.

The most important difference resides in the approximation involved in equation (7.22). Since h^2 for a given angle of scatter is large, the two-term approximation is only accurate at very low scattering angles: in the range $0.1°$ to $1.0°$.

It may thus be argued that the intensity of X-rays scattered at a small angle θ is given by

$$I_s = I_e(\theta) \left(\sum_i (f_i(h))^2 (1 - h^2 \langle s^2 \rangle/3) \right) \qquad (7.24)$$

where $\langle s^2 \rangle$ is the mean square radius of gyration, and $I_e(\theta)$ is the intensity scattered by a single electron in the direction θ.

It is conventional in X-ray work to replace $(1 - h^2 \langle s^2 \rangle/3)$ by $e^{(-h^2 \langle s^2 \rangle/3)}$, which is a valid approximation if $h^2 \langle s^2 \rangle/3$ is small, and then to take logarithms of both sides of equation (7.24) to obtain

$$\ln I_s = -h^2 \langle s^2 \rangle/3 + C \qquad (7.25)$$

where C is a constant containing all the other terms. This expression is termed the *Guinier expression* and it may be used to obtain estimates of the radius of gyration from plots of $\ln I_s$ versus $\sin^2 \frac{1}{2}\theta$.

The Guinier plot is only linear for small values of $\sin^2 \frac{1}{2}\theta$. For large values, deviations from linearity will occur that depend on the shape of the macromolecule in a complicated manner. We will only note that it is sometimes possible to obtain information concerning the shape - whether it be rod like, ellipsoidal etc. - from the analysis of such deviations.

If the molecules possess a regular repeating structure, such as is the case in a helical polymer, information can also be obtained concerning the repeat distances. Artificial order can sometimes be introduced into an otherwise chaotic assembly of solute molecules in solution (for instance by orientating rod-like molecules by flowing the solution through a narrow tube as mentioned in the next chapter), in order to improve the information that may be obtained concerning their shape. Ultimately, the solute molecules may be crystallised into a regular lattice and the angular dependence of the scattering analysed by the methods of X-ray crystallography.

7.10   Neutron scattering

The theory of elastic neutron scattering proceeds in a manner which is closely analogous to that of the scattering of electromagnetic radiation with but small modification.

These modifications must take into account the wavelength of the neutrons that are commonly employed and the differences in the scattering mechanism as indicated in section 7.2.

First, we should note that the wavelengths are large compared with the size of the nuclei, and so the scattering factor does not fall off with angle as for X-rays. In this respect neutron scattering is more akin to light scattering. However, since the wavelength is comparable with the distance between atoms in a molecule, the scattered intensity will vary with angle because of interference between the waves scattered from different atoms in the same molecule. For this reason neutron scattering may be used to measure radii of gyration of even small molecules. In this respect it is like X-ray scattering.

Secondly, we should note that the neutrons are not scattered by a dipole mechanism as is electromagnetic radiation. This means that the intensity of scattering is, in the absence of the interference effects mentioned above, isotropic and does not depend on the $1 + \cos^2 \theta$ factor discussed in section 7.3.

Finally, we must note that the neutron scattering factors of most common atoms (C, N, O) including deuterium are similar in magnitude, with only hydrogen showing a large difference. This means that the contrast (see section 7.7) will only be large if the macromolecule contains hydrogen atoms and solvent does not, or vice versa. For this reason it is usual to employ a solvent in which any hydrogen atoms have been replaced by deuterium atoms, D_2O, instead of H_2O for instance. If the macromolecule contains regions some of which are richer in hydrogen atoms than others (such as a nucleoprotein like a ribosomal particle or a nucleosome) it is possible to mask out one component or other by altering the proportion of D_2O to H_2O in the solvent until a match is achieved. When this is done the component whose scattering power is matched by the solvent will not scatter, and only that due to the other component will be observed.

It is also important to note that the radius of gyration that is measured will be that defined so that each scattering element is weighted according to the contrast it makes with the solvent. If the solvent is D_2O this will then correspond to the radius of gyration of the assembly of hydrogen atoms.

Finally, we emphasise that neutron scattering can provide information, as can X-ray scattering, concerning the shape and long range order in a macromolecule.

7.11 Laser-light scattering

Laser light can be thought of as a single plane-polarised monochromatic wave of high intensity extending over a long distance and continuing for a long period. This extension in space is termed *coherence length* and the extension in time is termed *temporal coherence*. The spatial and temporal coherence imply that the light has a very narrow bandwidth or that it has a high degree of monochromicity: there is a very small spread of wavelengths present.

A laser is a suitable light source for light-scattering measurements designed to obtain estimates of molecular weights and radii of gyration as explained above. Indeed its high intensity offers advantages in that only small scattering volumes are required so that less solution is needed and also because it is possible to make some allowance for the presence of dust particles as explained in section 7.6. However, the temporal coherence makes it possible to extract information concerning the Brownian motion of the macromolecule and so to measure the diffusion coefficients defined in the next chapter.

Let us first consider a dilute gas composed of point scatterers. These will be moving about with different speeds with a distribution of velocities determined by the Boltzmann law, and which is related to the diffusion coefficient (see section 8.12). A wave that is scattered by one of these moving particles suffers a change in frequency because of the Doppler effect. This means that the light scattered by the gas as a whole is distributed over a range of frequencies so that its spectrum mirrors the distribution of velocities. In principle the diffusion coefficient can be estimated from measurements of the manner in which the intensity of the scattered light varies with frequency, or from the width of the spectrum. It is important to note that the width of the spectrum is very small because the particles are moving at speeds that are small compared with the velocity of light. For this reason the broadening of the spectrum of the scattered light could only be detected if the incident beam were very closely monochromatic like that of a laser.

For solutions of macromolecules in solution there is again a broadening of the spectrum of the scattered light which is related to the diffusion coefficient, but in this case the origin cannot be properly attributed to Doppler shifts, and an alternative description of the effect must be sought.

In a liquid, the molecules collide with each other very frequently, maybe 10^{11} times per second. At each collision, the velocity of a molecule changes in a random way. This means that the wave trains scattered between collisions are of very short duration and cannot be represented by equation (7.1), which describes a wave of infinite duration. To describe a wave of such short duration, we must employ a sum of such terms and so describe a collection of waves with different frequencies. The vector sum of such a collection of waves, when they are chosen correctly, represents a wave of finite duration. We thus see that the frequent collisions have the effect of broadening the spectrum of the scattered light. These considerations also help to explain why the very narrow spectrum of the incident laser light requires a high degree of temporal coherence. We must now attempt to put these ideas on a quantitative basis.

Let us consider a single point scatterer. On account of its Brownian motion this will move about in the solution in a random manner. Let us suppose that at time zero it is at some starting point and that after a short interval of time, τ, it has moved to another point described by a vector s which denotes its displacement from its position at zero time. The field of the wave scattered to some point of observation at zero time may be described by (the real part of) $a\,e^{i\omega t}$. After the time interval, τ, the field will now be $a\,e^{i\omega t}\,e^{i(\omega\tau + h\cdot s)}$ where it has suffered a phase change $\omega\tau + h\cdot s$. The component $h\cdot s$ arises on account of the fact that it has moved to a new position and the component $\omega\tau$ because the time has changed. As time progresses this phase change will vary and, since the motion of the particle is random, the phase will vary in a random manner. Before proceeding let us note that if τ is zero the phase change is obviously zero and if τ is small the range of possible phase changes is small. As τ increases, the phase change becomes progressively more uncertain. This is to say, the correlation between the wave scattered at time zero and that at time τ decreases as τ increases.

A measure of this correlation is provided by the field *autocorrelation function*. This is defined as the average (as measured over a large number of separate time periods of duration τ) of the real part of the product $E^*(0)\,E(\tau)$ where $E^*(0)$ is the complex conjugate of the representation of the field at time zero and $E(\tau)$ is the representation of the field at time τ. If we suppose that $P(s_j, \tau)$ is the probability that the particle has suffered a displacement s_j during the time, τ, the autocorrelation function $G^1(\tau)$ may then be written

$$G^1(\tau) = \sum_j P(s_j, \tau)\, e^{i(\omega\tau + h\cdot s_j)} \tag{7.26}$$

If it is then assumed that the particle moves so as to satisfy the three-dimensional equivalent of Fick's second law (see section 8.12), and the summation over all possible displacements in equation (7.26) is replaced by an integration, it may be shown (see appendix C) that

$$G^1(\tau) = 2\, e^{-Dh^2\tau}\, e^{-i\omega\tau} \tag{7.27}$$

where D is the diffusion coefficient. It may be readily observed that the autocorrelation function is equal to unity for $\tau = 0$ and falls to zero as τ increases to infinity.

The next step in the argument is to note that the phase of the scattered wave fluctuates, as we have noted above, in a random manner, on account of the random motion of the particle. Because of this random element in the description of the scattered wave it cannot be represented as a simple sinusoidal wave. Instead it must be represented by the sum of a number of components whose resultant behaves in the manner of the scattered wave with its random phase. It turns out that this can only be done by taking a number of components with different frequencies. It also turns out that the spectrum of waves required, expressed as the intensity $I(\omega')$ of the components with angular frequency ω', is related by means of a mathematical identity to the autocorrelation function so that

$$I(\omega') = \frac{1}{2\pi} \int_0^\infty G^1(\tau)\, e^{i\omega'\tau}\, d\tau \tag{7.28}$$

This expression is sometimes referred to as the Weiner–Kinchine theorem. If we substitute equation (7.27) into this and take the real part we find that

$$I(\omega') = \frac{Dh^2/\pi}{(\omega' - \omega)^2 + (Dh^2)^2} \tag{7.29}$$

A spectrum whose shape is specified by an equation of the form of equation (7.29) is often termed a *Lorentzian* spectrum. We may see from equation (7.29) that $I(\omega')$ has its maximum value at a frequency corresponding to that of the incident wave at $\omega' = \omega$ and that this maximum value is $1/\pi Dh^2$. The value of the angular frequency when $I(\omega')$ is equal to half this maximum value is readily seen to be $\omega \pm Dh^2$, so that the width of the Lorentzian spectrum at half height is $\omega_{1/2} = 2Dh^2$. Thus, if the spectrum can be measured, D may be estimated given the value of $h(= 4\pi \sin \frac{1}{2}\theta/\lambda)$. It may be noted that if $D = 10^{-11}$ m^2/s and $\lambda = 546$ nm, then, for $\theta = 90°$, $\omega_{1/2} = 5297$ rad/s. Rather than attempt to measure such extremely narrow spectra by optical means, a different method is resorted to.

If two waves differing slightly in their frequency are allowed to interfere, they *beat* together. That is to say they combine together to give a resultant wave whose angular frequency is half the sum of the angular frequencies of the two components, ω_1 and ω_2, but whose amplitude varies in time with a frequency equal to $\frac{1}{2}(\omega_1 - \omega_2)$. This may be seen from the relation $\cos(\omega_1 t) + \cos(\omega_2 t) = 2 \cos\left[\frac{1}{2}(\omega_1 + \omega_2) t\right] \cos\left[\frac{1}{2}(\omega_1 - \omega_2) t\right]$. The intensity of this resultant varies in time also with a frequency $\frac{1}{2}(\omega_1 - \omega_2)$. Let us suppose that we mix with the scattered light a beam of frequency ω (this might be part of the incident beam deflected into the right direction by a mirror). It will combine with each element of frequency in the scattered light, ω', to produce a beat of angular frequency $\omega' - \omega$. This will be manifest in a periodic variation of the intensity of the combined wave in the audiofrequency region. This audiofrequency variation in the intensity is easily monitored and processed electronically. Since there is a whole range of closely spaced frequencies in the scattered light, these are converted into a range of audiofrequencies in the measured intensity. These audiofrequencies are easily analysed in an electronic frequency analyser to give an estimate of $\omega_{1/2}$. This is the so called heterodyne method in which the scattered light makes beats with an external beam. An alternative procedure is based on the fact that the different frequencies present in the scattered light can combine in a manner analogous to beating to give rise to fluctuations in the intensity, even in the absence of an external reference beam. These fluctuations can be monitored and analysed. It is perhaps worth noting that the quantity that is usually measured is the *intensity autocorrelation function* defined as

$$G^2(\tau) = \int_0^\infty I(t) I(t + \tau) \, \mathrm{d}t \qquad (7.30)$$

where $I(t)$ is the intensity measured at time t. Like the field autocorrelation function it is unity for $\tau = 0$ and falls to zero as τ increases. The intensity autocorrelation function is in turn related to the diffusion coefficient by

$$G^2(\tau) = 1 + a \, \mathrm{e}^{-2Dh^2\tau} \qquad (7.31)$$

so that D may be found from a plot of $\ln\{G^2(\tau) - 1\}$ versus τ. This method of employing the intensity autocorrelation function is often called *autocorrelation spectroscopy*. The method of employing the self-beating effect is often called the homodyne method or *self-beating spectroscopy*.

We have above presented a very schematic account of these methods for obtaining the diffusion coefficient of macromolecules. It has glossed over several features including the fact that the spectrum of the scattered light may also be broadened as a result of the rotational Brownian movement of the macromolecules. Thus, for asymmetric macromolecules, it is in principle possible to measure the rotational diffusion coefficients (see section 8.20).

In fact any motion of the macromolecule will give rise to changes in the spectrum of the scattered light. If, for instance, the macromolecule is moving with a steady translational velocity, the frequency of the scattered light will be changed by a fixed increment which is proportional to the velocity. If there are several different macromolecules moving with different velocities, the spectrum of the scattered light will manifest a number of Lorentzians shifted along the frequency axis by amounts proportional to their velocities and whose widths depend on their diffusion coefficients. This principle is used in the method of electrophoretic light scattering to measure the electrophoretic mobilities (see chapter 9) and diffusion coefficients of each component in a mixture of proteins.

There is little doubt that the application of these methods will result in other ingenious ways of studying macromolecules in solution.

Problems

7.1 A suspension of ribosomes of molecular weight 2.65×10^6 and radius of gyration 10 nm scatters light of wavelength 546 nm. The second virial coefficient, B, is 1.3×10^{-6} mol cm^3/g^2. Construct the Zimm plot using concentrations of 10, 20, 30, 40 and 50 mg/cm^3 and scattering angles of 20, 40, 60, 80 and 100°.

7.2 The intensity of X-rays scattered by a 50 mg/cm^3 protein solution falls by a factor of 10 when h increases from 0.1 to 1.0 nm^{-1}. If the Guinier plot is linear what is the molecular radius of gyration? How will the plot change when the concentration is increased?

8 The hydrodynamic properties of macromolecules

In previous chapters we have considered the properties of solutions of macromolecules at equilibrium such that these properties did not change with time. In this chapter we shall consider the rates at which various processes occur and how these rates depend on the conformation of the macromolecules or particles. More explicitly we shall consider the rates at which macromolecules move in the solution under the influence of various forces applied to them, but we shall also consider how such motion is manifest in changes in the observable properties of the solution itself.

The consideration of the response of macromolecules to applied forces will lead us to a consideration of the hydrodynamic equation of motion of a viscous fluid and thence to relations between the rates of such movement and the size and shape of the macromolecules, whereas the consideration of the relation of these movements to the observable properties will lead us to a consideration of the thermodynamics of irreversible processes. In this way we shall see that the measurement of various experimental parameters can lead to information concerning the size and shape and conformation of the macromolecules.

In all these considerations, the viscosity of the solvent in which the macromolecules are suspended is of the utmost importance. We shall begin with a discussion of this.

8.1 Viscosity

Consider first a small portion of some material, liquid or solid, in the form of a rectangular block with dimensions x, y and z as illustrated in figure 8.1. We suppose that the bottom face is fixed in space and that a force, F, is applied to the top face in the x direction. We suppose that this force acts uniformly over the area of the top face; as such it constitutes a shearing stress of magnitude F/xz.

This stress will tend to deform the material in the manner illustrated in figure 8.1 and produce a strain which is measured by the quantity $\Delta a/y$. This

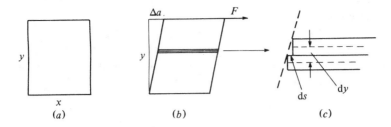

Figure 8.1. The rectangular block of material (a) is deformed by the force F as in (b). The shear strain is $\Delta a/y$. (c) shows a magnified portion of the deformed block. Two thin layers, separated by dy have moved a distance ds relative to each other. By similar triangles we see that ds/d$y = \Delta a/y$.

deformation can be looked upon as a sliding of the different layers of the material over one another.

If the material were an elastic solid, the existence of the strain would imply that the constituent parts of the material (atoms, molecules etc.) were displaced from the equilibrium positions dictated by the intermolecular forces acting between them. This displacement would lead to an increase in the energy (more exactly free energy) of the material and the generation of a restoring force which opposes the applied stress. At static equilibrium this restoring force exactly balances the applied force. If the material obeys Hooke's law, the restoring force is proportional to the strain, and the body is perfectly elastic. No real material, however, is perfectly elastic, and it is generally found that, for sufficiently large strains, the constituent parts of the material move into new positions of equilibrium. As this happens the restoring force will diminish or relax. If the applied stress is maintained constant, the strain will then continue to increase and the material will flow. This relaxation process takes place at a certain rate, which for a perfectly elastic solid is zero.

For a fluid such as a liquid, on the other hand, the relaxation process is rapid even for small strains and essentially no restoring force is manifest unless the different layers of the liquid are in relative motion. We thus see that a shearing stress applied to a liquid produces a continuously increasing strain. The rate of increase in strain in the example illustrated in figure 8.1 is measured by d$(\Delta a/y)$/dt, a quantity called the *shear rate*. If we consider two thin layers of the liquid separated by a distance dy we see from figure 8.1 and a consideration of similar triangles, that ds/d$y = \Delta a/y$, where ds is the distance moved by these two layers relative to each other. This implies that the shear rate may be written as d$(d$s$/d$y$)$/dt, which is equal to d$(d$s$/d$t$)$/dy. The quantity ds/dt is the relative velocity of the two layers, which we may write as u, so that the shear rate is du/dy. That is to say the shear rate is equal to the rate of change of velocity between the different layers or to the *velocity gradient*.

It is found in practice that the restoring force in a liquid increases as the shear rate increases so that at dynamic equilibrium it balances the applied stress. The ratio of the shear stress to the shear rate at dynamic equilibrium defines the *viscosity* of the liquid, η, so that

$$F/xz = \eta(du/dy) \tag{8.1}$$

The dimensions of viscosity are seen to be $ML^{-1}T^{-1}$ and the SI unit is kg/ms, one unit of which is equivalent to 10 poise, the unit in the older system of units. The viscosity of water at 20°C is 10^{-3} kg/ms.

If the viscosity of a liquid is independent of the shear rate, the liquid is said to be *Newtonian;* water is an example of a Newtonian fluid. For other, non-Newtonian, fluids the viscosity varies with the shear rate. Some materials show extreme forms of non-Newtonian behaviour and behave as elastic solids at low stresses and liquids at high stresses. It is important to note that the viscosity of most liquids is strongly dependent on temperature; for instance, the viscosity of water at room temperature decreases by 2.5% for each degree rise in temperature.

As the liquid flows under the influence of the applied stress, the latter does work and so supplies energy to the fluid. This energy is not stored in an elastic deformation as is the case with an elastic body. Instead the energy is dissipated so as to increase the random thermal motion of the constituent molecules of the liquid: that is to say it tends to increase the temperature. The energy dissipated per unit time in the portion of liquid referred to in figure 8.1 is clearly $F d(\Delta a)/dt$. Since $d(\Delta a/y)/dt = du/dt$, so that $d(\Delta a)/dt = y\, du/dy$, we see from equation (8.1) that the energy dissipated per unit time per unit volume is given by

$$(dU/dt)/xyz = \eta(du/dy)^2 \tag{8.2}$$

This relation shows that the viscosity is a measure of the rate of dissipation of energy per unit volume at unit shear rate.

We have so far considered a particularly simple situation in which the relative motion of adjacent layers of the fluid is along the x axis and varies only in the y direction. In more general situations the velocity of the fluid at any point must be described by a vector with three components. Each of these three components may vary in the two directions perpendicular to its direction to give six components of velocity gradient. Each of these six components gives rise to a component of shear stress and is related to it by equations similar to equation (8.1), and we may note here that the effect is mutual: if there is a component of velocity gradient in the moving fluid, there is a shear stress set up; and if there is a shear stress it causes there to be a velocity gradient. Similarly the total rate of energy dissipation per unit volume is equal to the sum of a series of terms similar to the right-hand side of equation (8.2). The modern description of these relations between shear stress and shear rate components requires the vector calculus and we shall not follow the argument further.

An important generalisation which is fully in accord with experiment concerns the layer of liquid in immediate contact with a solid object immersed in the liquid. This generalisation permits us to assume that this layer is always stationary; that is to say there is no slippage between the surface of the object and the layer of liquid in contact with it. Thus, if a macromolecule or particle is moving through a liquid or solvent, velocity gradients are necessarily set up in the surrounding fluid. The resulting shear stresses cause a resulting force to act on the particle opposing its motion and as energy is dissipated in the fluid the particle will come to rest unless there is an external force acting on it. These opposing forces constitute the *viscous drag* acting on the particle as a consequence of its motion.

8.2   Intrinsic viscosity

Let us consider a liquid in a state of uniform shear (as in figure 8.1). The velocity of the fluid increases uniformly in the y direction. Let us now suppose that a spherical particle is placed in this fluid as illustrated in figure 8.2. The velocity of the fluid at point A will be different from that at point B and this implies that the viscous drag at points A and B will be different. This means that a couple will act on the particle causing it to rotate. This rotational motion distorts the pattern of velocity gradients in the fluid and leads to a greater rate of dissipation of energy in the vicinity of the particle. Since the total rate of dissipation of energy is still given by equation (8.2), with du/dy still given by the overall velocity gradient, which we presume is maintained constant, it follows that the effective value of the viscosity of the fluid has been increased by the introduction of the spherical particle or particles. If the particle is not spherical, but asymmetric, similar effects will occur, but the particle will also tend to orientate in the fluid. The degree of orientation produced will depend on the magnitude of the shear gradient and on the tendency of Brownian motion to preserve random orientation. Thus the effective viscosity of the solution will still be greater than that of the pure solvent, but by an amount that will depend on the shear rate.

If η is the effective viscosity of the solution and η_0 that of the pure solvent, we define the *relative viscosity* of the solution by the ratio η/η_0 and the specific viscosity, η_{sp}, by $(\eta - \eta_0)/\eta_0$. It may be supposed that the more particles per unit volume we introduce into the solution, the greater will be the increased rate of dissipation of energy. We therefore go on to define the *reduced specific viscosity* by $(\eta - \eta_0)/\eta_0\rho_2$, where ρ_2 is the mass concentration of the macromolecular solute. The reduced specific viscosity of the solu-

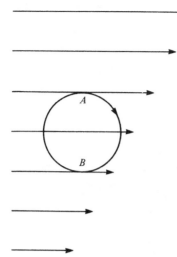

Figure 8.2. A spherical particle is immersed in a fluid in which there is a velocity gradient. The lengths of the arrows indicate the velocity of the fluid at different levels. The particle is carried along at a velocity equal to the velocity of the fluid at its centre, u_0. The fluid at A is moving faster than the particle and so exerts a viscous force. Similarly the fluid at B is moving slower than the particle and exerts a viscous force in the opposite direction. These two forces exert a couple which causes the particle to rotate in the direction shown.

tion is characteristic of the solute (in a particular solvent) except in so far as the particles are so close together that the distortion of the pattern of velocity gradients due to one particle influences those due to others. At infinite dilution we expect the effect due to the different particles present to be independent and therefore define the *intrinsic viscosity* as the limiting value of the reduced specific viscosity at infinite dilution. It then turns out that the intrinsic viscosity depends on the molecular weight, volume and shape of the particles constituting the macromolecular solute.

It should be noted that the relative and specific viscosities are dimensionless quantities (rather than having the dimensions of viscosity) and that the dimensions of intrinsic viscosity are those of reciprocal concentration.

The intrinsic viscosity of a macromolecular solute may be determined by measuring the viscosity of a series of solutions of different concentration and then plotting the reduced specific viscosity against the concentration as illustrated in figure 8.3. At low concentrations, the resulting plot is a straight line and the equation of this may be written in the form

$$\eta_{sp}/\rho_2 = [\eta] + k\rho_2 [\eta]^2 \tag{8.3}$$

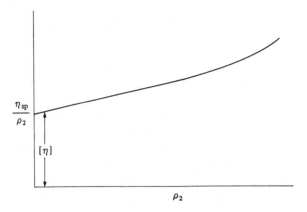

Figure 8.3. A plot of the reduced specific viscosity, η_{sp}/ρ_2, versus the concentration, ρ_2, is generally linear for small concentrations. The value of η_{sp}/ρ_2 extrapolated to zero concentration is the intrinsic viscosity, $[\eta]$. The slope of the linear portion of the plot is $k[\eta]^2$ where k is the Huggins constant.

and the constant k is known as the Huggins constant. This depends on the shape or conformation of the macromolecule.

It is sometimes convenient to eliminate the effect of molecular weight and size on the intrinsic viscosity by defining a further parameter, ν, by

$$[\eta] = \frac{N_A \nu v}{M_2} \qquad (8.4)$$

where M_2 is the molecular weight of the macromolecule and v its volume. ν then depends only on the shape or conformation.

We have so far neglected the fact mentioned above that the viscosity of a solution of asymmetric particles may depend on the shear rate. The ratio of the intrinsic viscosity at finite shear rate to that obtained by extrapolation to zero shear rate is found to depend on both the rotational frictional coefficients of the particle and the shear rate. We shall assume, unless stated to the contrary, the shear rate dependence is negligible or that the extrapolation to zero shear rate has been carried out in all references to intrinsic viscosity.

8.3 Frictional coefficients and frictional ratios

Let us suppose that a constant force, F, acts on a particle immersed in a solvent. This causes the particle to move, and in so doing a viscous drag is set up. This viscous drag turns out (for velocities not too large) to be proportional to the velocity of the particle, u. A state of dynamic equilibrium is reached when the force, F, is equal to the opposing viscous drag, and the particle moves with a steady terminal velocity. We then see that the terminal velocity is proportional to the constant force so that we may write

$$F = fu \qquad (8.5)$$

where the constant of proportionality, f, is known as the *translational frictional coefficient*.

The frictional coefficient turns out to be proportional to the viscosity of the surrounding solvent and otherwise to depend on the size and shape of the particle. The work done by the force, F, per unit time is clearly equal to Fu and hence is given by fu^2. The energy provided by the force in this manner is dissipated in the velocity gradient surrounding the particle.

It is convenient to define a further parameter, the frictional ratio, f_r, by

$$f_r = f/f_0 \qquad (8.6)$$

where f_0 is the frictional coefficient of a particle which is spherical in shape and has the same volume as the original particle. The point about the frictional ratio is that it depends only on the shape of the particle and not on its size.

It is important to note that the frictional coefficient of an asymmetric particle depends on the direction in which it is moving. Since, however, macromolecules are generally orientated at random in solution, it is usual to specify an average frictional coefficient (and frictional ratio) taken over all possible orientations with respect to the force, F.

The motion produced by a force in the manner just described is a translational motion. If, instead of a resultant force acting on the particle, a torque or couple acts, the particle will tend to rotate. In exactly the same way, a viscous drag is set up and a *rotational frictional coefficient* may be defined by

$$C = \zeta\omega \qquad (8.7)$$

where ω is the steady terminal angular velocity and C the couple acting. The rotational frictional coefficient ζ also turns out to be proportional to the viscosity of the solvent and otherwise to depend on the size and shape of the particle and we may define a rotational frictional ratio ζ_r by ζ/ζ_0, where ζ_0 is the rotational frictional coefficient of a spherical particle with the same volume. An asymmetrical particle will display a different rotational frictional coefficient according to the axis about which it is rotating, but, since any arbitrary rotation can be viewed as a simultaneous rotation about three orthogonal axes, we need specify only three rotational frictional coefficients (and ratios).

We have so far tacitly assumed that macromolecules undergoing translational or rotational motion are moving independently of each other. This implies that they are so far apart that the velocity gradients in the vicinity of one molecule do not interfere with those in the vicinity of another. This can only be so in dilute solution. At finite concentrations, the frictional coefficients will be concentration dependent.

We have also assumed that the translational or rotational motion is steady and uniform. For macromolecules in solution, this is a gross simplification. A macromolecule in solution suffers a continuous bombardment from solvent and other solute molecules. Every impact imparts energy to it and causes it to set off in random directions with random velocity, spinning with random angular velocity as it goes. The viscous forces acting on the particle will quickly nullify the effect of these random impulses and bring it to rest, only for it to suffer another impact. The particle will thus execute a random walk that constitutes its Brownian motion. Any motion due to external forces will be superimposed on this random walk and the work done per unit time by the external force will still be equal to Fu, where u is now the component of the instantaneous velocity that is due to the external force and proportional to it. Thus the work done is proportional to u^2 with a coefficient of proportionality that can be used to define the frictional coefficient. We are therefore justified in ignoring the background Brownian motion.

Two problems remain: first to relate the frictional coefficients to the size and shape of the macromolecules; this requires a consideration of the hydrodynamic equations of motion of the solvent. The second is to relate the frictional coefficients to measurable properties; this requires a consideration of irreversible thermodynamics.

8.4   The hydrodynamics of moving bodies

When a body or particle moves relative to the fluid surrounding it, the layer of fluid immediately adjacent to its surface moves, as we have indicated above, with it. This necessarily means that velocity gradients are set up in the surrounding fluid and the concomitant viscous forces exert a drag on the particle. In order to calculate this viscous drag it is necessary to obtain an expression for the magnitude and direction of the fluid velocity at all points in the vicinity of the particle.

This may be done in principle by considering the equation of motion of a small volume element of the fluid and equating the resultant of the forces acting on it to its rate of change of momentum. These forces include external forces (such as gravity), pressure forces and the viscous forces contingent on the local shear gradient. The resulting equation, known as the Navier–Stokes equation, then provides a set of partial differential equations whose solution, subject to the boundary conditions relevant to the problem in hand, must be sought. The Navier–Stokes equation is best expressed in terms of the vector calculus and for that reason we shall not quote it, still less attempt solutions.

In deriving solutions to the equations, certain simplifying assumptions are usually made. Thus the fluid is assumed to be isotropic, Newtonian, incompressible and of constant density. We also assume that the velocities involved are small so that the effects of inertia of the fluid element are negligible. Finally we seek time-independent steady-state solutions for which the pattern of flow in the vicinity of the particle is constant.

Even with these simplifying assumptions the solution of the Navier–Stokes equation is difficult, and exact solutions in closed form for the frictional coefficients or intrinsic viscosity have only been obtained for particles of simple geometric shape including ellipsoids of revolution.

Even in the case of ellipsoids of revolution the expressions obtained are only strictly valid for isolated particles and as such apply only to particles in solution at infinite dilution. At finite concentrations, the patterns of flow around the different particles interact, and these hydrodynamic interactions present grave theoretical difficulties. To circumvent these difficulties it is generally necessary to extrapolate measurements of frictional coefficients made at finite concentration to infinite dilution.

One final general problem remains to be considered. The use of the Navier–Stokes equation tacitly assumes that the fluid is a continuum or that it can be subdivided into smaller and smaller elements without limit and with no change in its properties. This clearly ignores the molecular structure of any real fluid (such as water). This assumption is justified in practice for bodies that are large compared with the molecules of the fluid but for particles the size of macromolecules is open to some doubt. Cheng and Schachman have measured the frictional coefficients of spherical particles (composed of polystyrene latex) of radii down to 260 nm and found them to be in good agreement with values calculated from their known size. The possibility remains, however, that the assumption is invalid for smaller particles.

8.5   The hydrodynamic properties of ellipsoids of revolution

An ellipsoid of revolution is the figure that is swept out when an ellipse is rotated about one of its axes, the axis of revolution. We suppose that the length of the axis of revolution is $2a$ and that the length of the other axis (the equatorial axis) is $2b$. The volume of the ellipsoid is then equal to $\frac{4}{3}\pi ab^2$ and its shape is defined by the axial ratio, $p = b/a$. If $p < 1$ the figure is a prolate (cigar shaped) ellipsoid, whereas if $p > 1$ it is an oblate (disc shaped) ellipsoid. If $p = 1$ the ellipsoid becomes a sphere. To avoid confusion, the reader should note that the axial ratio of an ellipsoid of revolution is defined by some authors to be the ratio of the major axis of the generating ellipse to its minor axis (and as such is greater than unity) and by others as the reciprocal of this.

The Navier–Stokes equation has been solved explicitly and exactly subject to the simplifying assumptions and relevant boundary conditions in the case of spherical and ellipsoidal particles.

Stokes first obtained the result of $6\pi\eta R$ for the translational frictional coefficient of a sphere of radius R and $6\eta v$ for the rotational frictional coefficient, where v is the volume, $\frac{4}{3}\pi R^3$. An expression for the intrinsic viscosity of a spherical particle was first obtained by Einstein who found that v in equation (8.4) is equal to $\frac{5}{2}$, giving $[\eta] = 5vN_A/2M_2$. In so far as the volume v is proportional to the mass of a sphere, it is noteworthy that the intrinsic viscosity is independent of the size of the sphere.

Perrin first obtained an expression for the translational frictional coefficient of an ellipsoid of revolution in terms of its volume and axial ratio. The resulting expression (averaged over all orientations of the ellipsoid) is given in table 8.1. He also obtained expressions for the rotational frictional coefficients (for rotation about the axis of revolution and the equatorial axis) as are also given in table 8.1. An expression for the intrinsic viscosity of ellipsoids was first given by Simha for the limiting case of zero shear rate in which the ellipsoids are orientated at random with respect to the direction of the gradient. This expression is also to be found in table 8.1. This result was later extended to finite shear rates by Scheraga who tabulated values of $[\eta]_g/[\eta]_0$ for different values of the shear gradient, g, and the axial ratio; the analytical expression for this ratio in terms of the parameters mentioned is of some complexity and is not given here.

The translational and rotational frictional ratios and the parameter v appearing in equation (8.4) are plotted against the axial ratio in figures 8.4.

8.6   The applicability of the hydrodynamics of ellipsoids of revolution

No macromolecule can be described accurately as an ellipsoid of revolution; even globular proteins which adopt compact conformations have an irregular shape as we have illustrated in figure 4.6. Nevertheless, it is often useful to suppose that their shape approximates to that of an ellipsoid and to interpret hydrodynamic measurements in terms of the volume and axial ratio of the *hydrodynamically equivalent ellipsoid of revolution;* that is to say the volume and axial ratio of the ellipsoid that has hydrodynamic properties identical to those of the real macromolecule. If the equivalent ellipsoid turns out to be very asymmetric, it may be supposed that the real particle is too. It should be borne in mind however that too close an identification of the shape of the real particle and the equivalent ellipsoid is unwarranted.

It must also be recognised that the hydrodynamic parameters that we have discussed depend on both the axial ratio and on the volume of the hydrodynamic particle. It must be emphasised that this volume must include not only the exclusion volume as discussed in section 3.3 but also the volume of any solvent that is tightly bound to the macromolecule or entrained within

Table 8.1. *Expressions for the hydrodynamic properties of an ellipsoid of revolution with semi-axis of revolution, a, and semi-equatorial axis, b*

For oblate or prolate ellipsoids of any axial ratio we have

$$f = 6\pi\eta r/p^{2/3}S$$

$$\zeta_b = 8\pi\eta r^3 \frac{2(1-p^2)}{3(1-p^2S)}$$

$$\zeta_a = 8\pi\eta r^3 \frac{2(1-p^4)}{3p[S(2-p^2)-1]}$$

$$\nu = \frac{2(1-p^2)^2}{15p^2}\left[\frac{\{2-41p^2+3p^2(8+5p^2)S\}}{\{2-5p^2+3p^4S\}\{(2+p)S-3\}}\right.$$

$$\left. +\frac{3\{1-2p^2+p^4S\}}{\{2(2-p^2)S-1\}\{1+2p^2-3p^4S\}}\right]$$

In the case of spheres, for which $p = 1$, these expressions become

$$f = 6\pi\eta r$$

$$\zeta = 8\pi\eta r^3$$

$$\nu = 5/2$$

In the case of prolate ellipsoids resembling rods for which $p \ll 1$ they become

$$f = 6\pi\eta a/\ln(2/p)$$

$$\zeta_b = \frac{16\pi\eta a^3}{3[2\ln(2a/b)-1]}$$

$$\nu = \frac{14}{15} + \frac{1}{5p^2[\ln(2/p)-\frac{1}{2}]}$$

f is the translational frictional coefficient; ζ_a is the rotational frictional coefficient for rotation of the equatorial axis about the axis of revolution; ζ_b is the rotational frictional coefficient for rotation of the axis of revolution about the equatorial axis. ν is the viscosity factor defined in equation (8.4). The parameters are given in terms of: p, the axial ratio defined as b/a; r, the radius of the sphere with the same volume as the ellipsoid and equal to $(ab^2)^{1/3}$; S which is defined for

Oblate ellipsoids $(p > 1)$ by $S = (p^2-1)^{-1/2}\tan^{-1}(p^2-1)^{1/2}$

Prolate ellipsoids $(p < 1)$ by $S = (1-p^2)^{-1/2}\ln\dfrac{1+(1-p^2)^{1/2}}{p}$

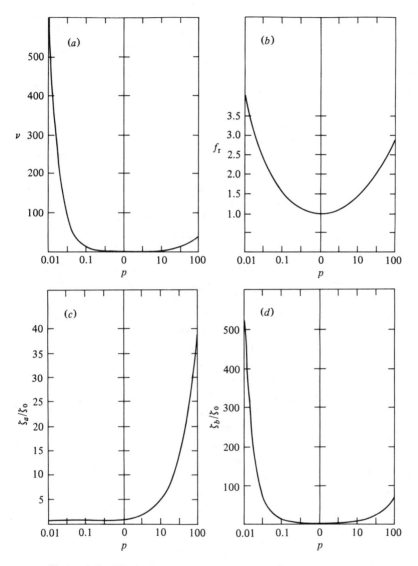

Figure 8.4. The hydrodynamic properties of ellipsoids of revolution. p is the axial ratio defined as the ratio of the equatorial axis to the axis of revolution. For prolate ellipsoids $p < 1$, and for oblate ellipsoids $p > 1$. For spheres $p = 1$. The parameters are plotted against p on a logarithmic scale.

 (a) The viscosity parameter, ν, appearing in equation (8.4).
 (b) The frictional ratio, f_r defined by equation (8.5).
 (c) The rotational frictional ratio ζ_a/ζ_0 for rotation of the equatorial axis about the axis of revolution.
 (d) The rotational frictional ratio ζ_b/ζ_0 for rotation of the axis of revolution about the equatorial axis.

it. Although some information concerning the amount of bound solvent can be obtained from studies of the mass of solvent that is bound to the dried protein as the partial pressure of the solvent is raised, the volume of the hydrodynamic particle cannot be measured with precision. It is invalid to equate the volume derived from the partial molar volume $(\overline{V}_2/N_A)$ to the hydrodynamic volume for reasons discussed in section 3.3 and because $\overline{V}_2/N_A$ certainly does not include a contribution for the volume of the bound solvent.

We shall see later in this chapter that in principle the volume terms may be eliminated from the expression for two hydrodynamic properties to give a measurable parameter that is independent of the volume of the particle and which depends only on the axial ratio; unfortunately the resulting expression does not vary greatly as the axial ratio is changed so that estimates of the latter in this way are subject to considerable error.

If the macromolecule is rod like, it is reasonable to suppose that it would behave like a prolate ellipsoid of revolution with small axial ratio. If the axial ratio is small the expressions given in table 8.1 may be simplified and the limiting expressions are also shown in this table. Numerical investigation also shows that, for small values of the axial ratio, the viscosity shape factor, ν, is approximately proportional to $p^{-1.8}$. If the length, L, of a rod-like molecule is equated to the length of the axis of revolution and the diameter to the length of the equatorial axis, $2b$, of a prolate ellipsoid, the axial ratio, p, of the rod is equal to $2b/L$. We therefore expect the intrinsic viscosity of a rod-like macromolecule to be proportional to L^{γ}, where the value of the exponent should be approximately equal to 1.8. Experimental values of this exponent usually fall in the range 1.0 to 1.8. It should be noted that, if the rod is to some degree flexible, the worm-like coil model must be invoked and that the exponent γ is expected to be less than 1.8.

Similar approximations can be made for the expression for the translational frictional ratio, and for small axial ratio we find that f_r is proportional to $L^{0.48}$. We shall return to the question of the hydrodynamic properties of rod-like macromolecules in section 8.8.

8.7   Random coils

An exact solution to the Navier–Stokes equation for an object with a geometry as complicated and as variable as that of a random coil presents insurmountable difficulties, and we must be content with approximate expressions.

First we shall summarise some experimental findings. Early work by Staudinger showed that a plot of the logarithm of the intrinsic viscosity of a series of homologous polymers (differing only in their molecular weight) against the logarithm of their molecular weight was frequently a straight line as illustrated in figure 8.5. This implies a relation of the form

$$[\eta] = kM^{\gamma} \tag{8.8}$$

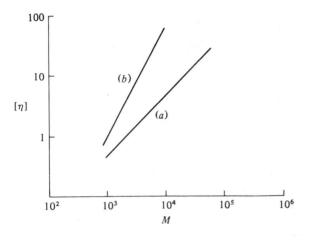

Figure 8.5. Typical double logarithmic plots of intrinsic viscosity versus molecular weight: (a) random-coil polymer at the theta temperature with $\gamma = 0.5$; (b) random coil in a good solvent with $\gamma = 0.8$.

For polymers in theta solvents, the exponent γ is frequently close to 0.5, whereas in good solvents it is larger and as high as 0.8. Later work showed that the translational frictional coefficient of random-coil molecules obeyed a similar law with an exponent close to 0.5 in theta solvents and somewhat less in good solvents. It is gratifying that these values of the exponents can be derived from the approximate theories that we describe below.

Several such approaches to the hydrodynamic properties of random coils have been descirbed. We shall content ourselves with discussing those based on the work of Kirkwood. He introduced the *pearl necklace model* of a polymer. In this, each monomer unit (or Kuhn statistical segment) is considered to behave as a small spherical particle of frictional coefficient f_b (of radius b, where $6\pi\eta b = f_b$). These particles or beads are supposed to be joined together by links which have negligible frictional resistance.

If the beads were sufficiently small, and far enough apart that the flow of solvent about one of them did not interfere with the flow round any of the others, the viscous drag on the whole chain would be merely the sum of the drags on all of the beads of σf_b, where σ is the number of beads in the chain. This situation represents the free draining limit and, if real polymers were to behave like this, the exponent, γ, relating frictional coefficient to molecular weight would be equal to unity.

The discrepancy between this free draining limit and experimental observation is to be traced to hydrodynamic interactions between the beads of the chain and in a realistic model these must be taken into account. Attempts to come to terms with this problem are quite involved and require considerable mathematical sophistication, and we will content ourselves with quoting the salient results of such investigations.

The results of Kirkwood's treatment expresses the translational frictional coefficient of a collection of beads, each of radius, b, in terms of the average of the reciprocal of the distance r_{ij} between the ith and the jth beads by

$$f = \frac{6\pi\eta b\sigma}{1 + \dfrac{b}{\sigma} \displaystyle\sum_{\substack{i=1 \\ i \neq j}}^{\sigma} \sum_{j=1}^{\sigma} \langle r_{ij}^{-1} \rangle} \tag{8.9}$$

The double summation extends over all values of i and j except for $i = j$ (i.e. for each pair of beads).

In order to evaluate this expression for a random coil, recourse must be had to an expression which defines the distribution of end-to-end distances of such a polymer. For unperturbed Gaussian coils, such an expression is provided by equation (4.25). In terms of this expression we see that the mean reciprocal end-to-end distance is given by $\int_0^1 r^{-1} \phi(r) \, dr / \int_0^\infty \phi(r) \, dr$. This evaluates to $(6/\langle r^2 \rangle_0 \pi)^{1/2}$ where $\langle r^2 \rangle_0$ is the mean square end-to-end distance of the unperturbed chain. If we take $|j - i| \beta^2$ to be the mean square distance between the ith and jth bead (as suggested by equation (4.18)) we see that $\langle r_{ij}^{-1} \rangle$ is equal to $(6/\pi |j - i| \beta^2)^{1/2}$. If this value of r_{ij}^{-1} is substituted into equation (8.9), and the summations replaced by integrals, there results

$$f = \frac{6\pi\eta\sigma b}{1 + \dfrac{b}{\beta} \sigma^{1/2} \dfrac{8}{3} \left(\dfrac{6}{\pi}\right)^{1/2}} \tag{8.10}$$

It must be emphasised that this expression is only expected to be valid for high-molecular-weight chains such that the Gaussian distribution of end-to-end distances is valid and such that the mean square end-to-end distance can be written in terms of the length of the Kuhn statistical segment, β, as $\sigma\beta^2$. Nevertheless, if σ is small, such that $\sigma^{1/2} \ll \beta/b$, equation (8.10) reduces to the expression appropriate to the free draining coil as discussed above.

Of more interest is the limit for large σ such that $\sigma^{1/2} \gg \beta/b$. In this case equation (8.10) reduces after replacing $\sigma\beta^2$ by $6\langle s^2 \rangle_0$ to

$$f = 6\pi\eta \langle s^2 \rangle_0^{1/2} \tfrac{3}{8}\pi^{1/2} \tag{8.11}$$

This shows that high-molecular-weight random coils are expected to behave like solid spherical particles with a radius equal to $0.665 \ (= 3\pi^{1/2}/8)$ times the root mean square radius of gyration of the unperturbed coil. In this limit the coil is said to be non-draining or impermeable.

Similar reasoning may be applied to derive expressions for the rotational frictional coefficient and the intrinsic viscosity. Thus, in the non-draining limits, we have

$$\zeta = 8\pi\eta (0.89 \langle s^2 \rangle_0^{1/2})^3 \tag{8.12}$$

$$[\eta] = \frac{5}{2}\frac{N_A}{M_2}\frac{4}{3}\pi(0.87\langle s^2\rangle_0^{1/2})^3 \tag{8.13}$$

In each case we see that the unperturbed coil behaves like a solid sphere with a radius equal to a constant times the root mean square radius of gyration, though the constant is different for the three different types of hydrodynamic property.

We have seen in Chapter 4 that the mean square radius of gyration of the unperturbed coil may be expressed as $\langle s^2\rangle_0 = \sigma\beta^2/6$. This implies that f, ζ and $[\eta]$ are proportional to $M^{0.5}$, $M^{1.5}$ and $M^{0.5}$ respectively for a series of homologous polymers of molecular weight M. These predictions have been amply verified in the case of the translational frictional coefficient and the intrinsic viscosity for several polymer-solvent systems. Indeed the molecular-weight dependences specified by these relations appear even to hold at molecular weights too low for the non-draining limiting expression to be valid.

It is also found by experiment that the ratio, $f/6\pi\eta\langle s^2\rangle_0^{1/2}$ is close to the theoretical value of 0.665, though the value of the ratio $[\eta] M_2/N_A\langle s^2\rangle_0^{3/2}$ is found to be somewhat less than the predicted value of 6.90 given by equation (8.13).

The physical significance of the impermeable coil in the non-draining limit deserves further comment. If the coil obeys Gaussian statistics, the density of segments is much greater at points closer to the centre of mass than at the periphery, so that one may envisage that solvent flows more or less freely past the peripheral beads, but is prevented from flowing through the centre. This implies that solvent at the centre is carried along with the moving coil. It should not however be thought that this solvent is in any thermodynamic sense bound to the polymer molecule; the effect is solely a consequence of hydrodynamic interactions.

The validity of equations (8.11), (8.12) and (8.13) depends on the assumption that the end-to-end distances are distributed in a Gaussian manner for the evaluation of the terms $\langle r_{ij}^{-1}\rangle$ in equation (8.9). This is reasonable for unperturbed coils in theta solvent for which $\langle r^2\rangle_0 = \sigma\beta^2$. In highly expanded coils in good solvents, however, we have $\langle r^2\rangle = \sigma\beta^2\alpha^2$, and the expansion factor α itself depends on the chain length as discussed in section 6.5. Furthermore, the coil becomes non-Gaussian and in order to take this effect into account a more complex analysis is required, which has been provided by several authors. Even if we were to ignore this effect and assume the validity of equations (8.11), (8.12) and (8.13), the dependence of the hydrodynamic parameters on molecular weight would be different from that suggested above. In good solvents we have suggested in section 6.5 that α is proportional to $M^{0.1}$. This implies on the basis of these three equations that f, ζ and $[\eta]$ should be proportional to $M^{0.6}$, $M^{1.8}$ and $M^{0.8}$ respectively if the unperturbed radius of gyration $\langle s^2\rangle_0$ is replaced by the expression proportional to

$M^{0.6}$, which is appropriate to a good solvent. It is noteworthy that f and $[\eta]$ in good solvents do in fact depend on the molecular weight in this manner, but that the values of the constants $f/6\pi\eta\langle s^2\rangle^{1/2}$ and $[\eta]\,M_2/N_A\langle s^2\rangle^{3/2}$ both differ from their values in theta solvents and also depend on the solvent (or on α).

8.8   Rod-like molecules and molecules with other shapes

The result of the Kirkwood theory embodied in equation (8.9) is valid for any assembly of spherical beads and is not limited to random coils. It can therefore be used to calculate the translational frictional coefficient of a particle of any shape provided it may be assumed that it behaves like a collection of spherical beads.

For such a collection of beads, arranged in definite spatial positions, the process of averaging over all possible conformations is redundant and all that is required is the summation of the distance between the centres of all pairs of beads. The process of averaging over all orientations of a rigid particle is already implied by equation (8.9).

Two methods for modelling a rigid particle of arbitrary shape in terms of spherical particles have been proposed. In the first of these the shape of a convex body (i.e. one without reentrant crevices etc.) is approximated by placing a large number of small beads of identical size all over its surface. It is then assumed that the resulting hollow shell behaves hydrodynamically like the particle in question. The sum of the reciprocal distances between each pair of beads must then be calculated in order to obtain the frictional coefficient from equation (8.9). If the particle is of arbitrary shape and a large number of beads is used this can be a tedious process, since for n beads there are $n^2 - n$ reciprocal distances to be determined.

This hollow-shell method has been applied to a spherical particle and gave a result that was identical to that obtained by Stokes when the number of infinitesimally small beads was taken to the limit of infinity. It might be thought that such a modelling process can never be exact in so far as there must always be small spaces between small spheres in contact; it is therefore interesting that the proportional error introduced when a fraction, y, of the beads were left out of the shell so as to leave gaps was only of the order of $\frac{1}{4}y$. This procedure is nevertheless not exact, for when it was applied to prolate ellipsoids of revolution significant differences between the results obtained and the exact results derived from Perrin's formula were apparent. The discrepancy can be traced to approximations inherent in equation (8.9).

One of the limitations of equation (8.9) as it stands is the requirement that the beads be all of the same size. This limitation may be removed by an application of the hollow-shell model to obtain an expression that applies to a collection of spherical subunits of arbitrary size. To obtain this expression each subunit of radius b_i is modelled by a shell composed of beads of identical radius. The resulting expression is

$$f = \frac{6\pi\eta \left(\sum_{i=1}^{n} b_i^2 \right)^2}{\sum_{i=1}^{n} b_i^3 + \sum_{i=1}^{n} \sum_{\substack{j=1 \\ i \neq j}}^{n} b_i^2 b_j^2 r_{ij}^{-1}} \tag{8.14}$$

where n is the total number of subunits and r_{ij} is the distance between the centres of subunit i and j. If all the b_i are identical, this reduces to equation (8.9).

It is equation (8.14) that is used in the second method of modelling particles of arbitrary shape. In this method the particle is approximated by a relatively small number of spherical particles in contact with each other such that the envelope of the resulting assembly resembles that of the particle in question. This procedure is illustrated in figure 8.6 for a rod and for a prolate ellipsoid.

In the case of cylindrical rods we may define the axial ratio of a rod of length, L, and diameter, d, to be $p = d/L$. Application of the method then gives an approximate expression for the translational frictional coefficient valid if n, the number of spheres in the model ($= 1/p$), is not too small in the form

$$f = \frac{1}{p} \frac{3\pi\eta \, d}{[\frac{1}{2}p - \ln p + 0.577]} \tag{8.15}$$

This expression is similar to more exact expression derived in other ways and may also be compared with the expression given in table 8.1 for the translational frictional coefficient of prolate ellipsoids of revolution in the limit of large axial ratio.

The method has also been applied to particles such as bacteriophage which have very complicated shapes with reasonably satisfactory result. It should be borne in mind, however, that such results are not exact on account of errors inherent in equation (8.9), which result from approximations made in taking account of the hydrodynamic interactions between the beads. More recently attempts have been made to remove these approximations, but the application of these results requires a digital computer to solve a large number of equations in order to obtain a numerical result.

Finally, we note that equation (8.9) has been applied to the worm-like coil to obtain an expression for the translation frictional coefficient in terms of the persistence length. A similar expression has been obtained for the intrinsic viscosity.

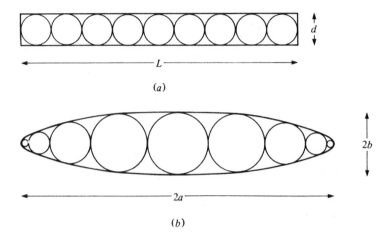

(a)

(b)

Figure 8.6. (a) The form of a cylindrical rod of length, L, and diameter, d, is approximated by L/d touching spheres of diameter, d. (b) The form of a prolate ellipsoid of revolution is approximated by nine touching spheres of diameter graded such that they are tangential to the surface of the ellipsoid.

8.9   The measurement of intrinsic viscosity

So far in this chapter we have discussed the relation between the frictional coefficients and the intrinsic viscosity, on the one hand, and the size and shape of the macromolecular particle on the other. We must now discuss the relation between the hydrodynamic parameters and experimentally accessible quantities.

The measurement of the intrinsic viscosity presents no new problems. All that is required is the measurement of the specific viscosity of several solutions of the macromolecule at known concentrations, and the extrapolation of the reduced specific viscosity to zero concentration as indicated in figure 8.3. The viscosity of the solutions may be measured in a capillary viscometer. In this a well-defined volume of solution is allowed to flow slowly through a narrow vertically suspended capillary tube. The time required for this to occur may be shown to be proportional to η/ρ, where η is the viscosity of the solution and ρ its density. The reduced specific viscosity is then given by

$$\eta_{sp}/\rho_2 = \frac{(\eta - \eta_0)}{\eta_0 \rho_2} = \frac{(t - t_0)}{t_0 \rho_2} + \left(\frac{1}{\rho_0} - \bar{v}_2\right)\frac{t}{t_0} \qquad (8.16)$$

where t and t_0 are the flow times for a solution of weight concentration ρ_2 and for solvent. ρ_0 is the density of the pure solvent and $\bar{v}_2$ the partial specific volume of the macromolecular solute. The term in $1/\rho_0 - \bar{v}_2$ corrects for the effect of the solute on the density of the solution.

A major disadvantage of the capillary viscometer is that the shear rate is ill defined and variable. This disadvantage is overcome in the Couette viscometer in which the solution whose viscosity is to be measured is placed in the annular gap of thickness d between two concentric cylinders. One of these is rotated at a constant angular velocity, ω, and the torque, C, transmitted to the other is measured. For narrow gaps between cylinders of radius, R, the shear rate is equal to $R\omega/d$ and the viscosity equal to $dC/2\pi L\omega R^3$, where L is the length of the cylinders.

We have noted in section 8.1 that the viscosity of a liquid may vary rapidly with temperature; for this reason it is necessary to ensure that the viscometer and its contents are well thermostatted. We have also noted that the dimensions of intrinsic viscosity are those of reciprocal concentration. It has been conventional to define the intrinsic viscosity as the pure number resulting when the concentration is measured in units of grams per decilitre; for this reason a factor of 100 occurs in formulae involving the intrinsic viscosity in most books. For reasons of consistency we shall not follow this convention, but continue to assume that the dimensions of $[\eta]$ are $L^3 M^{-1}$ so that the factors of 100 do not appear.

8.10   Flow rates and the equation of continuity

Frictional coefficients could be measured directly were it possible to measure both the force acting on a macromolecule and the velocity it produced. Unfortunately macromolecules are mostly too small for the velocities to be measured directly. All that is possible is to measure changes in the concentration at various points in the vessel containing the solution as macromolecules move within it. This leads us to a consideration of the flows of the various components present and the relation of the flows to the forces that produce them.

For simplicity we shall only consider one-dimensional flow in the x direction and assume that the concentrations of the components are constant in the y and z directions. The relations we shall obtain are readily extended to three-dimensional flow, but we shall not pursue such complexities.

The *flow rate* or *flux*, J_i, of component i in a solution is defined as the net amount of that component which passes in unit time through a plane of unit area and which is normal to the direction of movement (i.e. in the yz plane). We shall only consider fluxes defined in terms of the mass of the component, but some authors prefer them defined in terms of the number of moles.

The flux, J_i, at any point in the solution, is easily related to the mean velocity of the molecules of the component at that point, u_i. Thus, in time

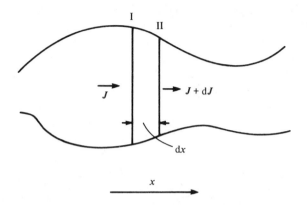

Figure 8.7. The cross-sectional area of the thin slice of solution at I is A; that at II is $A + dA$. The flow rate at I is J and that at II is $J + dJ$. The thickness of the slice is dx so that its volume is $A\,dx$.

dt, every molecule originally situated between two planes of area A (both normal to the direction of motion) and separated by a distance $u_i\,dt$, will pass through the downstream plane. The volume between these planes is $u_i A\,dt$ so that the mass of the component therein is $u_i \rho_i A\,dt$. It thus follows that

$$J_i = u_i \rho_i \tag{8.17}$$

If the flux varies in the direction of movement, the mass of component flowing in across the upstream plane will not be the same as that flowing out across the downstream plane and the component will tend to accululate (or disperse from) the volume element between the planes. This implies that the concentration there will change. Let us suppose that the fluxes at the positions of two planes separated by a distance dx and of areas A and $A + dA$ are J_i and $J_i + dJ_i$ as illustrated in figure 8.7. The mass entering across the first plane will be $J_i A\,dt$ and that leaving across the second plane will be $(J_i + dJ_i)(A + dA_i)\,dt$. The difference between these two masses divided by the volume between the planes, $A\,dx$, gives the increase in concentration in time dt so that

$$\left(\frac{\partial \rho_i}{\partial t}\right)_x = -\frac{1}{A}\left(\frac{\partial (JA)}{\partial\,dx}\right)_t \tag{8.18}$$

This equation is referred to as the *equation of continuity* and is based on the principle of the conservation of mass. It implies that, in a vessel of constant cross-sectional area, the concentration at any point can only change if there is a gradient in the concentration or the velocity of the component in question.

So far we have tacitly assumed that the plane used to define the fluxes is stationary with respect to the vessel. This is a natural assumption and the most convenient from the viewpoint of an experiment in which we measure the rates of change of concentration at fixed points in the vessel. It is, however, not the only possible assumption and it turns out that the theory of flow processes is simpler if the reference plane moves with the local centre of mass or the local centre of volume. The velocity of the local centre of volume u^* is defined by the relation

$$u^* = \sum_i \bar{v}_i \rho_i u_i \qquad (8.19)$$

where u_i is the velocity of the component i relative to the vessel. Fluxes defined for this frame of reference are then given by $(u_i - u^*) \rho_i$, and it is easy to show that $\sum_i (u_i - u^*) \bar{v}_i \rho_i = 0$. This means that the fluxes of the different components present are not all independent of each other.

It also turns out that if the partial specific volumes, $\bar{v}_i$, of the different components are constant (and hence independent of concentration or any other relevant factors), the velocity $u^* = 0$ and fluxes relative to the centre of volume are equal to fluxes relative to the vessel. We shall assume, except where stated, that this is the case.

The fluxes that we have defined above correspond to translational flow rates, but the concept of a flux can be generalised to describe the rates of other processes that occur in systems that are not at thermodynamic equilibrium. Thus we could define a flux of heat as the rate at which energy is transported as heat across a plane, or a flux corresponding to a chemical reaction as the rate at which products are formed from reactants per unit volume. More relevant to the present discussion are rotational flows. Let us suppose that the macromolecules present in a solution have an axis which for a particular molecule at a given time points in a given direction making an angle ϕ with, say, the x direction. We may define a rotational concentration $\rho_i(\phi)$ such that $\rho_i(\phi) \, d\phi$ is the mass of all macromolecules in a unit volume such that the molecular axis makes an angle with the x axis lying in the range ϕ to $\phi + d\phi$. If the molecules are orientated at random we may suppose that $\rho_i(\phi)$ is independent of ϕ so that $d\rho_i(\phi)/d\phi = 0$ but, if this is not the case, a rotational concentration gradient, $d\rho_i(\phi)/d\phi$, exists. We may now define a rotational flux as the mass of all molecules present in unit volume whose axis rotates through the value ϕ in unit time in the direction of increasing ϕ. This rotational flux, $J_i(\phi)$, will in general vary with ϕ and is a consequence of torques applied to the macromolecules. The rotational fluxes are related to the angular velocities of the macromolecules by an expression analogous to equation (8.17) and an equation of continuity analogous to equation (8.18) can also be derived.

Finally, it is worth noting that the fluxes discussed above are essentially macroscopic properties of the solution; that is to say, their definition in no way presupposes any knowledge of arrangement of the various molecules in the solution. Viewed in this light, equation (8.17) may serve as a definition of the mean velocity, u_i, of a component.

8.11 The origin of the flows

The translational flows or fluxes discussed in the previous section are a result of the forces acting on the various components present. It is thus clear that the effect of an external force, such as centrifugal force in a centrifuge, is to cause movement of the constituents of the solution. In this context it is relevant to note that, as a molecule moves in the direction of the force acting, its potential energy Φ_i decreases. It is this decrease in potential energy that is dissipated by the viscous forces operating. We may also note that the gradient of the potential energy, $d\Phi_i/dx$, is equal to $-F_i$, where F_i is the magnitude of the force acting in the x direction.

We should now note that external forces are not the only factors that may give rise to flows. Thus if the chemical potential of a component varies from one part of the vessel to another such that there is a gradient of chemical potential, $d\mu_i/dx$, flow will occur. Just as a gradient of mechanical potential energy is equivalent to a force, so is a gradient of chemical potential. We shall see in the next section that this, perhaps mysterious, result follows naturally from an examination of the situation in the light of irreversible thermodynamics. Meanwhile the following argument may make the proposition more reasonable.

First, we note that a gradient of chemical potential implies (in the situation where there are no other external forces acting) a concentration gradient. Let us consider a vessel of constant cross-sectional area A as illustrated in figure 8.8. The vessel is filled with a solution in which there is a concentration $d\rho_2/dx$ of solute at position 0. We next consider two thin slices of the solution, I and II, on either side of 0, and each of thickness Δx. The mean concentration in I is $\rho_2 - \frac{1}{2}(d\rho_2/dx)\,\Delta x$ and that in II is $\rho_2 + \frac{1}{2}(d\rho_2/dx)\,\Delta x$. We now note that the solute molecules are moving about at random as they collide with solvent and other solute molecules, an effect identifiable with their Brownian motion. Let us suppose that the average distance moved by a given molecule in time Δt is Δx, which means that of all the molecules in I, half will have moved a distance Δx in the direction of 0 in time Δt and half will have moved the same distance away from 0. It is thus clear that a mass of solute equal to $\frac{1}{2}\{\rho_2 - \frac{1}{2}(d\rho_2/dx)\,\Delta x\}\,A\,dx$ will have moved from I into II. Similarly a mass $\frac{1}{2}\{\rho + \frac{1}{2}(d\rho_2/dx)\,\Delta x\}\,A\,dx$ will have moved from II into I. The difference between these two quantities is not zero and so there is a net flux of solute across the plane at 0. This flux is clearly seen to be $J_2 = \frac{1}{4}(d\rho_2/dx)\,(\Delta x)^2$. We thus see that a flux is a natural consequence of a concentration gradient. This process, whereby components tend to flow down

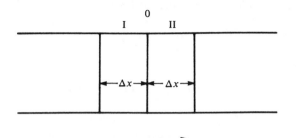

Figure 8.8. The cross-sectional area of the channel is A. The thickness of both elements I and II is Δx, equal to the mean distance moved by a molecule in time, t, either to the right or to the left. The concentration increases uniformly to the right so that $d\rho/dx = g$. The concentration at the centre of element I is $\rho - \frac{1}{2}g\Delta x$ and that at the centre of element II is $\rho + \frac{1}{2}g\Delta x$. In time, t, half the molecules in element I move to the right into element II and half those in element II move to the left into element I.

concentration gradients, is termed *diffusion*. In an analogous manner it may be argued that rotational concentration gradients as defined in the previous section give rise to rotational flows by *rotational diffusion*. In section 8.13 we will consider translational diffusion in more detail, but before that we must pay closer attention to the specification of the forces acting through a consideration of irreversible thermodynamics.

8.12   Irreversible thermodynamics

Classical thermodynamics, the subject matter of chapter 3, deals only with systems at equilibrium. Here we are concerned with systems which are not at equilibrium and in which the processes occur at rates described by the various fluxes or flows. These are zero at equilibrium.

One approach to classical thermodynamics casts the second law in the form

$$\Delta S = \Delta S_e + \Delta S_i \qquad (8.20)$$

where ΔS is the total entropy change in a system during some process; ΔS_e is the change in entropy resulting from the exchange of heat or matter with its surroundings across its boundaries and ΔS_i is the entropy created within the system as a result of irreversible processes taking place therein. If no irreversible processes are taking place inside the system, $\Delta S_i = 0$, but otherwise $\Delta S_i > 0$. Thus, for a system in which various irreversible processes are taking place, we may define a rate of increase of entropy per unit volume as

$$\theta = \frac{1}{V}\frac{dS_i}{dt} \qquad (8.21)$$

The fundamental postulate of irreversible thermodynamics is that each flux, J_i, is associated with a quantity X_i such that

$$T\theta = \sum_i J_i X_i \tag{8.22}$$

The quantities X_i are variously referred to as *thermodynamic forces* or *affinities*. They may be viewed as causing the fluxes.

From a consideration of the relations that hold between the relevant thermodynamic variables it may thus be shown that a flow of heat, for instance, is associated with a temperature gradient; a translational flow of some component is associated with a gradient in its mechanical or chemical potential; a rotational flow is associated with a torque or a rotational concentration gradient.

A subtlety arises when we note that there is some latitude allowed in the definition of the fluxes. For instance we have noted in section 8.10 that translational flows may be defined in different ways according to the reference plane that we assume. Since the rate of entropy production cannot depend on the manner in which we define the fluxes, a change in definition of the fluxes must result in a change in the form of the forces in order that equation (8.22) remains true. Thus if J^α and X^α are the fluxes and forces defined in one way and J^β and X^β those defined in another, equation (8.22) implies that $\Sigma J^\alpha X^\alpha = \Sigma J^\beta X^\beta$ and so provides a recipe for deriving the forces relevant to one definition of the flows given the expressions relevant to another.

The first step therefore in the application of irreversible thermodynamics is to derive an expression of the rate of production of entropy. This will be found to consist of a sum of terms, one for each irreversible process as implied by equation (8.22). We then attempt to factorise each term into a J and an X component and identify the fluxes and forces with these factors. Unfortunately this cannot be done unambiguously for, even if we were to find one pair of factors, J and X, which gave a product JX, another pair $J\alpha$ and $\alpha^{-1} X$ would give the same product irrespective of α. To complete the analysis we require another relation between the flows and forces. This is provided by the *phenomenological equations* which postulate a relation between the forces and fluxes of the form

$$J_i = \sum_j L_{ij} X_j \tag{8.23}$$

That is to say we postulate linear relations between each J and every X. Two points should be noted. Firstly, we allow for the possibility that each of the forces is instrumental in bringing about any of the fluxes. For instance a flow of one component may in part be due to a gradient in its own concentration and in part due to gradients in the concentration of all the other components present. That is to say that the flows of the different components may be coupled.

Secondly, we have postulated that the relation between flows and forces is linear. Provided that the system is not too far removed from equilibrium, arguments may be adduced to show that this is a reasonable approximation. It is believed to be valid in the context of the experimental situations discussed below. Nevertheless, if the system is very far from equilibrium, equation (8.23) is no longer valid and other, more complicated, non-linear expressions must be used instead. Such situations prevail in living cells, and the application of the non-linear theory can provide interesting insights into the possible modes of physicochemical behaviour manifest in living organisms. It would deflect us too far from the subject of this book, however, to pursue this line of thought further.

The constants, L_{ij}, which appear in equation (8.23) are known as the *phenomenological coefficients*. If there are n processes, giving rise to n fluxes and n forces, there are n^2 coefficients. However, they are not all independent, for it may be shown that, provided the fluxes and forces are properly chosen so as to comply with equation 8.22,

$$L_{ij} = L_{ji} \qquad (8.24)$$

These relations, known as the *Onsager reciprocal relations*, imply there are only $\frac{1}{2}n(n+1)$ independent phenomenological coefficients.

Finally, we may note that, if the fluxes are chosen (for instance in the manner discussed in section 8.11) in such a way that they are not all independent, then the forces are not all independent either. If we so chose the fluxes, then we may eliminate from consideration the flux and force corresponding to the flow of one of the components, usually the solvent, component 1.

Let us now consider the application of these procedures to the one-dimensional flow along the x direction in a solution of n components. Let us suppose that the external force acting on unit mass of component i is F_i, and that the mass concentration and chemical potential at any point in the vessel of component i are ρ_i and μ_i respectively; we also suppose that the concentration gradient of i at any point, $d\rho_i/dx$, is g_i, and that the entire solution is at the same temperature so that there is no flow of heat. We shall also suppose that the flows, J_i, of each component are defined to be relative to a plane moving with the local centre of volume as discussed in section 8.10. We shall finally suppose that F_i, ρ_i, μ_i, g_i and J_i may all vary with position, x, in the cell or vessel containing the solution.

By considering the rate of production of entropy in a small volume element of the solution it is possible to derive the exact and general result that the force associated with the flow of any component i (except that of the solvent) is given by

$$X_i = F_i - \frac{\bar{v}_i}{\bar{v}_1} F_1 - \sum_{j=2}^{n} \frac{1}{M_i} \frac{\partial \mu_i}{\partial \rho_j} g_j - \frac{\bar{v}_i}{\bar{v}_1} \sum_{j=2}^{n} \sum_{k=2}^{n} \frac{1}{M_k} \frac{\rho_k \partial \mu_k}{\rho_1 \partial \rho_j} g_j \qquad (8.25)$$

A schematic account of the derivation of this expression is given in appendix D. Here we only note that it is derived using only standard thermodynamic relations and equation (8.22).

If there are only two components present, a solvent and a solute, considerable simplification results, and there is only one force, X_2, associated with the flow of solute to be considered. It is given by

$$X_2 = F_2 - \frac{\bar{v}_2}{\bar{v}_1} F_1 - \frac{1}{M_2 \bar{v}_1 \rho_1} \frac{\partial \mu_2}{\partial \rho_2} g_2 \qquad (8.26)$$

Similarly there is only one phenomenological equation and one phenomenological coefficient so that

$$J_2 = L_{22} X_2 \qquad (8.27)$$

The reader should note the occurrence in equation (8.25) of terms of the form $\partial \mu_i / \partial \rho_j$ which specify the change in the chemical potential of component i which occurs when the concentration of component j is altered, temperature, pressure and the concentration of the other components remaining constant.

Finally it may be noted that a more extended version of equation (8.25) may be derived, which includes terms relevant to the rotation of macromolecules under the influence of external torques and to rotational diffusion, but we will not dwell on the details of this extension.

8.13   Frictional coefficients and the phenomenological coefficients

In section 8.2 we defined the frictional coefficient of a particle as the ratio of the force acting on a particle (at infinite dilution) to the velocity produced by it. We must now consider the relation between the frictional coefficient and the phenomenological coefficients. We will do this explicitly only for a two-component system.

First, we note that the term X_2 in equation (8.26) represents the force acting on unit mass of component 2 so that $M_2 X_2 / N_A$ represents the force on one molecule.

Next, we point out that the definition of the velocity of the particle is determined by the boundary conditions employed in the solution of the Navier–Stokes equation. Since the movement of the particles in one direction through the solvent implies a net flow of the solvent displaced in the opposite direction, it is appropriate to define the velocity of the particles to be used in the calculation of the frictional coefficient to be $u_2 - u_1$, the velocity of the particles relative to the solvent. The boundary conditions must be chosen in such a manner to be consistent with this. It should be noted that, with this definition, the velocity u_1 falls to zero at infinite dilution when the expressions for the frictional coefficients discussed in a previous section of this chapter apply.

We next note that the flux of solute, J_2, relative to the centre of volume is equal to $\rho_2(u_2 - u^*)$ and that this may readily be shown to be equal to $\rho_1\rho_2\bar{v}_1(u_2 - u_1)$ for a two-component system.

It then follows that

$$f = \frac{M_2 X_2}{N_A(u_2 - u_1)} = \frac{M_2\rho_1\rho_2\bar{v}_1}{N_A L_{22}} \qquad (8.28)$$

where we have eliminated the ratio X_2/J_2 by the use of equation (8.27).

In equation (8.28), f is the translational frictional coefficient of the solute at finite concentration in a two-component solution. In principle (though not in practice) it may be calculated in terms of the size and shape of the macromolecule, provided appropriate boundary conditions are employed in the solution of the Navier-Stokes equation and that hydrodynamic interactions are taken into account. For multicomponent systems additional complexities arise in that interactions between the different solute particles must be considered; again, in principle, these may be related to the phenomenological coefficients L_{ij} $(i \neq j)$, but we will not consider this difficult matter further.

8.14 Diffusion

We have argued in section 8.11 that gradients in chemical potentials can act as forces and give rise to translational flows. We are now in a position to put these ideas on a quantitative footing. This we do by considering the situation in which there are no external forces so that $F_i = 0$. We then see from equation (8.26) for a two-component solution that

$$X_2 = -(\partial\mu_2/\partial\rho_2)\, g_2/\bar{v}_1\rho_1 M_2 \qquad (8.29)$$

and hence from equation (8.27) that there is a flow of solute given by

$$J_2 = -L_{22}(\partial\mu_2/\partial\rho_2)\, g_2/\bar{v}_1\rho_1 M_2 \qquad (8.30)$$

A relation of the form of equation (8.29) was anticipated by Fick who proposed that

$$J_2 = -D_2 g_2 \qquad (8.31)$$

This relation, known as Fick's first law of diffusion, defines the *diffusion coefficient*, D_2. A comparison of equations (8.30) and (8.31) and the expansion of the term $\partial\mu_2/\partial\rho_2$ with the aid of equation (3.23) leads to an expression for the diffusion coefficient:

$$D_2 = \frac{RTL_{22}}{M_2 v_1 \rho_1 \rho_2} [1 + \rho_2 (\partial \ln y_2/\partial\rho_2)] \qquad (8.32)$$

where y_2 is the activity coefficient of the solute on the molar concentration scale. For dilute solutions, $\partial \ln y_2/\partial\rho_2$ is almost zero.

The elimination of L_{22} from equations (8.32) and (8.28) then leads to

$$D_2 = \frac{RT}{N_A f} [1 + \rho_2 (\partial \ln y_2/\partial\rho_2)] \qquad (8.33)$$

We thus see that f can be calculated from experimental values of the diffusion coefficient.

The term in y_2 in equation (8.33) represents the thermodynamic effect of finite concentration on the diffusion coefficient. An additional hydrodynamic effect of finite concentration is implicit in the concentration dependence of the frictional coefficient, f. Only at infinite dilution can this be equated to one of the expressions for the frictional coefficient discussed in earlier sections.

At infinite dilution equation (8.33) reduces to

$$D_2^0 = RT/N_A f^0 \qquad (8.34)$$

This relation was first derived by Einstein on the basis of kinetic theory and the superscript 0 denotes the value pertaining to infinite dilution.

So far we have only considered two-component solutions. When there is more than one solute component, relations of greater complexity ensue. These may be expressed, for a three-component system, as an example, by

$$J_2 = -D_{22}g_2 - D_{23}g_3$$

$$J_3 = -D_{32}g_2 - D_{33}g_3 \qquad (8.35)$$

The Onsager reciprocal relations imply a relation (which we shall not quote but which has been experimentally verified) between D_{22}, D_{33} and the cross diffusion coefficients D_{23} and D_{32}. The cross diffusion coefficients describe the coupling of the flow of one solute component to concentration gradients of the other, and result partly from thermodynamic effects through the mediation of the terms $\partial\mu_i/\partial\rho_j$ appearing in equation (8.25) and partly from hydrodynamic interactions and their effect on the phenomenological coefficients.

8.15　The measurement of diffusion coefficients

Several techniques have been employed for the measurements of diffusion coefficients, one of which we have already encountered in chapter 7. First we may note that diffusion coefficients have the dimensions of $L^2 T^{-1}$

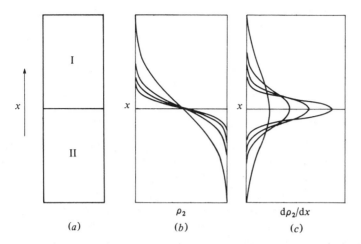

ρ_2 $d\rho_2/dx$

(a) (b) (c)

Figure 8.9. Free boundary diffusion. (a) The lower region of the vessel, II, is filled with solution of concentration ρ_0. The upper region, I, is filled with solvent to form an initially sharp boundary at the origin. (b) As time increases the concentration gradient at the origin decreases as the boundary spreads owing to diffusion. (c) The 'peaks' in the concentration gradient profile become broader and flatter as time progresses.

and for macromolecules are of the order of 10^{-11} m^2/s or 10^{-7} cm^2/s. A unit of 10^{-11} m^2/s is sometimes used and named a *Fick*.

If we apply the equation of continuity, equation (8.18), to equation (8.31) there results an expression known as Fick's second law of diffusion which may be cast in the form

$$(\partial\rho_2/\partial t)_x = D_2(\partial^2\rho_2/\partial x^2) \qquad (8.36)$$

This is a partial differential equation whose solution, subject to appropriate boundary conditions, specifies the manner in which the concentration of the solute depends on time and position in the cell.

In the method of free boundary diffusion we arrange for a rectangular vessel to be half filled with a solution containing the solute whose diffusion coefficient is to be measured. On top of this solution a layer of solvent is carefully added so as to form an initially sharp interface with the solution. There is thus a large concentration gradient at the interface. This causes diffusion to occur across the interface or boundary and the boundary becomes diffuse as illustrated in figure 8.9. As time progresses the boundary becomes more and more diffuse till at infinite time the concentration is uniform throughout the vessel and the concentration gradient is reduced to zero at all points. We assume that the diffusion coefficient is independent of concentration and solve equation (8.36) subject to the conditions that at $t = 0$, $\rho_2 = 0$ for $x > 0$ and $\rho_2 = \rho_0$ for $x < 0$, where ρ_0 is the initial concentration of the solution below the interface and x is a coordinate denoting position up and down the cell with $x = 0$ at the interface. We may thus show that

$$\left(\frac{\partial \rho_2}{\partial x}\right)_t = \frac{-\rho_0}{(u\pi Dt)^{1/2}} e^{-x^2/4Dt} \tag{8.37}$$

$$\rho_2 = \frac{\rho_0}{2}\left[1 - \frac{2}{\pi^{1/2}} \int_0^{x/(4Dt)^{1/2}} e^{-y^2} dy\right] \tag{8.38}$$

These equations specify the shape of the concentration versus x and concentration gradient versus x curves illustrated in figure 8.9 and the manner in which they depend on time.

With the aid of various optical devices, either the concentration or concentration gradient profiles may be monitored and, from such curves, estimates of the diffusion coefficient may be obtained. One method requires the concentration gradient curves, which we may note are Gaussian in shape and symmetrical about $x = 0$. From equation (8.37) we see that the height of the curve at $x = 0$, h_{max}, is equal to $\rho_0/(4\pi Dt)^{1/2}$ and it may also be shown that the area between the curve and the x axis, A, is constant and equal to ρ_0. We thus see that $A/h_{max} = (4\pi Dt)^{1/2}$ so that a plot of $(A/h_{max})^2$ versus t, for curves obtained at various times, is linear with a slope equal to $4\pi D$. It may be noted that if the diffusion coefficient is appreciably dependent on the concentration, the Gaussian curves are distorted and no longer symmetric. Similarly if several solute components with different diffusion coefficients are present the curves will generally be flattened.

It is usually necessary to perform several such experiments at different concentrations, ρ_0, and extrapolate the apparent values of the diffusion coefficient so obtained to zero concentration to obtain D^0.

It is important to note that, since the frictional coefficient depends on the viscosity of the solvent, the temperature must be maintained constant throughout the time of the experiment, which may last for several days for a high-molecular-weight solute. It may be noted that this disadvantage of long times for high-molecular-weight solutes is not present in the method based on light scattering discussed in chapter 7; indeed, the light-scattering method is easier for high-molecular-weight solutes, which scatter more light. In this sense the two methods are complementary.

Finally, it should be noted that it is common practice to correct measured diffusion coefficients to refer to a standard temperature and standard solvent. This correction is performed by a relation which takes account of the dependence of D on the temperature and viscosity of the solvent band which is implied by equation (8.34):

$$D_s^0 = \frac{T_s}{T_e} \frac{\eta_e}{\eta_s} D_e^0 \tag{8.39}$$

where subscript s denotes parameters relevant to the standard conditions and subscript e to the experimental conditions. It should be noted that diffusion coefficients corrected in this way are not necessarily the same as would be obtained if they were actually measured under standard conditions. This is obviously the case if the conformations of the macromolecule are different in the two solvents.

Later we shall see that diffusion coefficients may also be obtained by methods analogous to free boundary methods but conducted in the ultracentrifuge.

8.16 Sedimentation in the ultracentrifuge

A suspension of macroscopic particles which have a higher density than the solvent will sediment to the bottom of the vessel containing them. That is to say they move downwards through the solution under the influence of the force of gravity (conversely if they have a lower density they float to the surface). As they sediment down they leave behind them in the upper part of the vessel a region free of particles and consisting of pure solvent. There is thus a boundary set up between a region inhabited by particles and a region free of them. The rate at which this boundary and the particles themselves descend decreases as the particles get smaller. For particles as small as macromolecules the rate is so small that the sedimentation is effectively counteracted by diffusion due to the concentration gradient at the boundary. The sedimentation of such macromolecules will only occur if the force acting on the particles is artificially increased by spinning the solution at high speed in an ultracentrifuge. Under these circumstances the force of gravity is replaced by a centrifugal force.

In a modern ultracentrifuge the solution is contained in a small cell held in a rotor which spins at angular velocities of up to 70 000 rev/min. The centripetal acceleration $\omega^2 x$ of the cell at a radial distance x may then approach 3×10^8 cm/s^2 as opposed to the acceleration due to gravity of only 10^3 cm/s^2. Since the rate of sedimentation depends on the frictional coefficient, and hence on the viscosity, arrangements are made to maintain the temperature of the cell constant. Arrangements are also made by the use of various optical devices (similar to those used in free boundary diffusion experiments) to monitor the concentration of the solute or its concentration gradient as a function of radial distance.

The cell itself is so constructed that the solution is held in a cavity between two quartz plates which allow the passage of light required by the monitoring optical system. The sides of this cavity are arranged to lie along radii through the centre of rotation of the rotor as illustrated in figure 8.10. If the sides of the cell were parallel, a sedimenting molecule moving along a radius would, if close to the side, strike the side and cause convective disturbances. This is

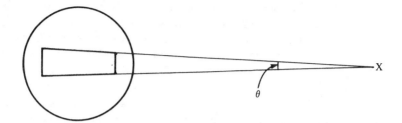

Figure 8.10. Diagram of ultracentrifuge cell. The sides of the sector-shaped cavity lie along radii that meet at the centre of rotation, X. The angle subtended by the sides of the cell to the centre is θ.

obviated by making the cell sector shaped as described. This means that the cross-sectional area of the cell increases as the distance, x, from the centre increases according to

$$A = \theta h x \qquad (8.40)$$

where h is the thickness of the cell between the two quartz windows, and θ is the angle subtended to the centre (see figure 8.10).

There are many different sorts of ingenious experiments that may be conducted using the ultracentrifuge. In one common experiment, designed to obtain information concerning the rate of sedimentation, the cell is filled to a certain height (measured from the 'bottom' of the cell at the point nearest the circumference and situated at a radial distance x_b) with a solution of uniform concentration c_0. A small air space is left at the top of the cell to form an air–solution interface at a radial distance x_m.

As time progresses the macromolecules sediment towards the circumference of the rotor or the bottom of the cell causing a boundary to form between two regions of the cell (the so called plateau regions) of uniform concentration. This boundary tends to spread, as the experiment continues, owing to diffusion (see previous section) and the boundary moves down the cell. The general manner in which the concentration and concentration gradient change with time and radial distance is illustrated in figure 8.11. Such an experiment is termed a sedimentation-velocity experiment.

8.17   Flow rates in the ultracentrifuge

We must now consider quantitatively the fluxes that occur in an ultracentrifuge cell. We have already noted that the centrifugal force acting on unit mass at a radial distance, x, is $\omega^2 x$, where ω is the angular velocity.

Let us first consider the forces acting on a particle of mass m and volume v. These are twofold: first, there is the centrifugal force $\omega^2 x m$ acting on the particle itself; secondly, there is a buoyancy force $\omega^2 x v \rho$ acting in the oppo-

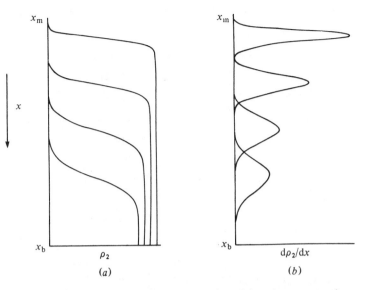

Figure 8.11. Concentration profiles (a) and concentration-gradient profiles (b) in the ultracentrifuge cell at various times during a sedimentation-velocity experiment. The concentration is initially uniform between the meniscus at radial distance x_m and the cell bottom x_b. The boundary between solvent and solution moves down the cell as time progresses; at the same time the concentration gradient at the boundary decreases.

site direction. Thus the resultant force acting in a centrifugal direction is $\omega^2 x(m - v\rho)$. This force quickly accelerates the particle (in a time roughly of the order of 10^{-10} s) to a steady terminal velocity u_2 relative to the cell which is determined by the frictional coefficient so that $\omega^2 x(m - v\rho) = fu_2$. The term $m - v\rho$ may be written as $M_2(1 - v\rho/m)/N_A$, where v/m is the volume per unit mass of the solute particles, which it is tempting to equate to the partial specific volume of the solute. We should at this stage of the argument note that this identification of v/m with $\bar{v}_2$ is speculative and that uncertainty shrouds the precise definition of the density ρ; is it the density of the solvent or that of the solution?

These matters are clarified and resolved by the application of irreversible thermodynamics. To do this we first note that F_i (for all components i), the external force per unit mass, should be put equal to $\omega^2 x$. For a two-component solution it then follows from equation (8.26) that

$$X_2 = \omega^2 x(1 - \bar{v}_2 \rho)/\bar{v}_1 \rho_1 - (\partial\mu_2/\partial\rho_2) g_2/\bar{v}_1 \rho_1 M_2 \qquad (8.41)$$

We may now define the *sedimentation coefficient*, s_2, of the solute as the ratio of the velocity, relative to the cell, of the solute to the centrifugal acceleration under such circumstances that the external force is the only force acting (i.e., $g_2 = 0$) so that

$$s_2 = (u_2)_{g_2 = 0}/\omega^2 x \tag{8.42}$$

Since u_2 is defined to be relative to the cell, s_2 is, in principle, measurable.

Let us now consider the flow, J_2, defined as relative to the centre of volume so that $J_2 = \rho_2(u_2 - u^*)$ and suppose that the partial specific volume, $\bar{v}_2$, is independent of concentration so that $u^* = 0$ as explained in section 8.10. It then follows from equations (8.27) and (8.41) that

$$J_2 = \rho_2 u_2 = \frac{L_{22}}{\bar{v}_1 \rho_1} \left[\omega^2 x (1 - \bar{v}_2 \rho) - \frac{\partial \mu_2}{\partial \rho_2} \frac{g_2}{M_2} \right] \tag{8.43}$$

If we now put $g_2 = 0$ we may replace $u_2/\omega^2 x$ by s_2 and thus obtain

$$L_{22} = \frac{s_2 \rho_2 \bar{v}_1 \rho_1}{(1 - v_1 \rho_1)} \tag{8.44}$$

Finally L_{22} may be expressed in terms of f, the frictional coefficient, with the aid of equation (8.28) to give

$$s_2 = \frac{M_2 (1 - \bar{v}_2 \rho)}{N_A f} \tag{8.45}$$

Comparing this with the simple expression derived above we see that the term v/m may indeed be replaced by $\bar{v}_2$ and also that the factor ρ is unambiguously the density of the solution. We may also observe that the sedimentation coefficient is expressed by equation (8.45) in terms of properties of the solute, M_2, $\bar{v}_2$ and f so that the frictional coefficient may be estimated from experimental determinations of the sedimentation coefficient, provided M_2 and $\bar{v}_2$ are also known.

L_{22} is the phenomenological coefficient and appears in equations (8.32) and (8.44); its value is the same in both equations. If we eliminate L_{22} between them there results a relation which in the limit of infinite dilution is

$$M_2 = RTs_2^0/D_2^0(1 - \bar{v}_2 \rho) \tag{8.46}$$

This relation is known as the *Svedberg equation*, and it clearly provides a method of estimating the molecular weight of a macromolecule from measurements of both its sedimentation and diffusion coefficients, if both are extrapolated to infinite dilution.

To complete our analysis of sedimentation of two-component solutions we may eliminate L_{22} from equation (8.43) by means of equations (8.44) and (8.32) to obtain

$$J_2 = \omega^2 x s_2 \rho_2 - D_2 g_2 \tag{8.47}$$

If sedimentation in multicomponent systems is analysed in a similar fashion, and we note that the sedimentation coefficient of component i may always be defined as $u_i/\omega^2 x$, there result expressions of greater complexity. For ternary solutions, for instance, expressions analogous to equation (8.47) may be derived which are of the form

$$J_2 = \omega^2 x s_2 \rho_2 - D_{22} g_2 - D_{23} g_3$$

$$J_3 = \omega^2 x s_3 \rho_3 - D_{32} g_2 - D_{33} g_3 \tag{8.48}$$

and in which the coefficients, s_2, s_3, D_{22}, D_{23}, D_{33} and D_{32} are interrelated on account of the Onsager reciprocal relations. Similarly an analogue of the Svedberg equation (8.46) may be derived, but this is of some complexity and will not be quoted.

8.18   The Lamm equation

If the equation of continuity is applied to equation (8.47) there results the Lamm equation written

$$\left(\frac{\partial \rho_2}{\partial t}\right)_x = -\frac{1}{x}\frac{\partial}{\partial x}\left(s\omega^2 x^2 \rho_2 - Dx\frac{\partial \rho_2}{\partial x}\right)_t \tag{8.49}$$

In deriving this result we have assumed that the cross-sectional area of the cell is proportional to x as stated in equation (8.40).

The Lamm equation is a partial differential equation whose solution gives the concentration as a function of time and radial distance in a manner which depends on the boundary conditions appropriate for the experiment. For a sedimentation-velocity experiment the appropriate boundary conditions are, at $t = 0$, $\rho_2 = \rho_0$ for $x_m < x < x_b$ and for all t, $\rho_2 = 0$ for $x_m > x > x_b$.

The solution of the equation is not simple and is rendered even more difficult if either the sedimentation or the diffusion coefficients are appreciably concentration dependent. Nevertheless, solutions involving various approximations have been obtained, and these show that the concentration or concentration gradient follows the form illustrated in figure 8.11. From such plots it is possible to derive accurate estimates of the sedimentation coefficient and, under favourable circumstances, of the diffusion coefficient as well. It is important to note that the sedimentation coefficients so derived generally show a dependence on the concentration of the solute and that in multicomponent systems the estimates are complicated by interactions between the flows as implied by equation (8.48).

An interesting situation prevails at the meniscus and at the bottom of the cell, for at these points there can be no flow of solute. If J_2 is set equal to zero in equation (8.47) we obtain the relation

$$\frac{s_2}{D_2} = \frac{1}{\omega^2 x \rho_2}\frac{\partial \rho_2}{\partial x} \tag{8.50}$$

The ratio s_2/D_2 is related by the Svedberg equation to the molar mass so that we see that measurement of $(\partial\rho_2/\partial x)/x\rho_2$ at either the meniscus or bottom of the cell will yield an estimate of the molar mass. This method of estimating molecular weights is referred to as the Archibald method, after its inventor.

8.19   The measurement of sedimentation coefficients

Associated with the boundary between solution and solvent that moves down the cell in a sedimentation-velocity experiment is a 'peak' in the concentration gradient profile as illustrated in figure 8.11. Approximate values of the sedimentation coefficient may be obtained by measuring the radial distance, x_h, of the top of such peaks (where $\partial^2\rho_2/\partial x^2 = 0$) and plotting $\ln x_h$ against time. The slope of the resulting plot may be shown to be equal to $s\omega^2$. If the 'peak' is broad and the boundary diffuse a more detailed analysis of the shape of the concentration gradient profile is necessitated and x_h must be redefined by

$$x_h^2 = \frac{\displaystyle\int_{x_m}^{x_b} x^2 \left(\frac{\partial\rho_2}{\partial x}\right) dx}{\displaystyle\int_{x_m}^{x_b} \left(\frac{\partial\rho_2}{\partial x}\right) dx} \tag{8.51}$$

The sedimentation coefficient as defined by equation (8.42) has the dimensions T and is characteristically of the order of 10^{-13} s; a conventional unit called the *Svedberg unit* (S) equal to 10^{-13} s is usually employed in the specification of sedimentation coefficients.

The sedimentation coefficient is generally found to depend on the concentration and generally to decrease as the concentration increases. For this reason it is desirable to extrapolate experimental sedimentation coefficients to zero concentration using a relation of the form $s = s^0/(1 + kc)$, where k is some constant and s^0 is the sedimentation coefficient at zero concentration.

The sedimentation coefficient depends on the density and viscosity of the solvent as is apparent from equation (8.45). For this reason it is conventional to 'correct' measured values to refer to a standard solvent at a standard temperature as is done for diffusion. Thus using the same subscript notation as employed in equation (8.39)

$$s_s^0 = s_e^0 \frac{(1 - \bar{v}_{2,s}\rho_s)}{(1 - \bar{v}_{2,e}\rho_e)} \frac{\eta_e}{\eta_s} \tag{8.52}$$

Finally, we remark that, if the frictional coefficient for a series of homologous polymers is proportional to M^γ, the sedimentation coefficient is proportional to $M^{1-\gamma}$.

8.20   Sedimentation equilibrium

If an experiment in the ultracentrifuge is continued for a very long time a state of equilibrium will be approached in which no further changes will be manifest, and the flow rate J_2 will be zero throughout the cell. The concentration will then be well defined at all radial positions in the cell. The nature of this equilibrium concentration distribution can be obtained directly from equation (8.43), but it is instructive to obtain the identical result by using only the methods of equilibrium thermodynamics.

First we note that the pressure, P, increases as the radial distance, x, increases. This increase in pressure is analogous to the increase in atmospheric pressure as the altitude decreases. Consider the forces acting on an element of the solution at radial distance, x, of thickness, dx, and cross sectional area, dA. There is a downwards force $P\,dA$ and an upwards force $\{P + (dP/dx)\,dx\}\,dA$ acting on this element as illustrated in figure 8.12. The difference between these two forces is a result of the centrifugal force $\rho\omega^2 x\,dA\,dx$ acting on the element. At equilibrium the net force must be zero so that

$$\frac{dP}{dx} = \omega^2 x\rho \tag{8.53}$$

where ρ is the density of the solution in the element.

The general conditions for thermodynamic equilibrium for such a system are that the temperature must be uniform throughout the cell and the net energy change for the transfer of dn_i moles of component i from one position (at radial distance x) to another (at radial distance x') is zero. This energy change is compounded of two contributions; the first is equal to $\mu_{i,x'}\,dn_i - \mu_{i,x}\,dn_i$ being the change in Gibbs free energy associated with such a change and where $\mu_{i,x'}$ and $\mu_{i,x}$ are the chemical potentials at the two positions. The second contribution is the difference of potential energy in the centrifugal field $M_i\Phi_{x'}\,dn_i - M_i\Phi_x\,dn_i$, where Φ_x is the potential energy of a unit mass at position x. If we equate the sum of these two differences to zero we see that

$$(\mu_{i,x'} + M_2\Phi_{x'})\,dn_i = (\mu_{i,x} + M_i\Phi_x)\,dn_i \tag{8.54}$$

Since this must be true whatever the magnitude of dn_i and whichever two positions, x' and x, we select, the quantity $\mu_{i,x} + M_i\Phi_x$ must be a constant throughout the cell for each component. This quantity is sometimes referred to as the total chemical potential. If we differentiate the total chemical potential with respect to x we then see that

$$\frac{d\mu_i}{dx} = M_i\omega^2 x \tag{8.55}$$

where we have replaced the gradient of potential energy per unit mass by

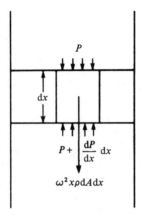

Figure 8.12. Hydrostatic equilibrium for an element of solution of cross-sectional area dA, and thickness, dx. The pressure on the upper face is P and that on the lower face is $P + (dP/dx)\, dx$. The net force on the element due to this pressure differential is balanced by the centrifugal force acting downwards and equal to $\omega^2 x\rho\, dA\, dx$.

$-\omega^2 x$, the force per unit mass. The minus sign arises because the energy decreases as x increases down the cell.

Now the chemical potential of a component varies from position to position in the cell for two reasons: because of the change of pressure therein and because of the change in concentration. For a two-component system we may thus write as a mathematical identity

$$\frac{d\mu_2}{dx} = \left(\frac{\partial\mu_2}{\partial\rho_2}\right)_{T,\rho} \frac{d\rho_2}{dx} + \left(\frac{\partial\mu_2}{\partial P}\right)_{T,\rho_2} \frac{dP}{dx} \qquad (8.56)$$

We may then show from equations (8.53), (8.55) and (8.56), again expanding $\partial\mu_2/\partial\rho_2$ by equation (3.23) and noting that $\partial\mu_2/\partial P$ is equal to $M_2\bar{v}_2$ that

$$M_2\omega^2 x(1 - \bar{v}_2\rho) = \frac{RT}{\rho_2} \frac{d\rho_2}{dx} \left[1 + \rho_2\left(\frac{\partial \ln y_2}{\partial\rho_2}\right)_{T,P}\right] \qquad (8.57)$$

This relation may be used to estimate the molecular weight of the solute from measurements of the concentration or concentration-gradient profiles that appertain to the equilibrium distribution of solute in the ultracentrifuge cell. It is usually necessary to obtain several estimates of the apparent molecular weight, using different initial solute concentrations, and to extrapolate to zero concentration in order to eliminate the effect of the activity coefficient, y_2, at finite concentrations.

A similar but more complicated expression holds for each component in a multicomponent mixture and, in such cases, it is possible to obtain estimates of the number-average, weight-average or z-average molecular weights, according to the manner in which the experimental data are analysed. It should be emphasised that great care must be taken if non-ideality effects are also present, particularly if the chemical potential of one macromolecular species depends on the concentration of the others.

8.21 Rotational diffusion coefficients and relaxation times

The formalism of irreversible thermodynamics may also be applied to the rotational motion of macromolecules to obtain expressions for the rotational flow rates or fluxes discussed in section 8.11. We will not analyse such situations in detail save to point out that, if there is a rotational concentration gradient, there will result a rotational flow to which it is related by a *rotational diffusion coefficient*. In general for an asymmetric particle there will be three rotational diffusion coefficients corresponding to rotation of the three axes of the molecule. These rotational diffusion coefficients are in turn related to the rotational frictional coefficients by equations analogous to equation (8.34)

$$\Theta^0 = RT/N_A \zeta^0 \tag{8.58}$$

Closely allied to rotational diffusion coefficients are rotational *relaxation times;* these quantities are pertinent to the experimental investigation of the rotational motion of macromolecules. Let us suppose that all the molecules in a solution are aligned at zero time so that a given axis on every molecule points in the same direction. Let ϕ denote the angle between the axis of a given molecule and this direction. There is then, at $t = 0$, an infinite rotational concentration gradient at $\phi = 0$. As time progresses rotational diffusion due to Brownian motion will cause the molecules to become disorientated. We may measure the degree of orientation left at any future time by the average value of $\cos \phi$, $\langle \cos \phi \rangle$, taken over all molecules present. At zero time, $\langle \cos \phi \rangle = 1$ and, when the molecules are completely disorganised and orientated at random, $\langle \cos \phi \rangle = 0$. Detailed analysis shows that $\langle \cos \phi \rangle$ falls exponentially with time, t, according to

$$\langle \cos \phi \rangle = e^{-t/\tau} \tag{8.59}$$

where τ is termed the *rotational relaxation time* of the axis in question. The process of disorientation of the direction of the axis in question results from rotational diffusion about the other two axes and is related to the rotational diffusion coefficients about these other two axes by

$$\tau_1 = 1/(\Theta_2 + \Theta_3) \tag{8.60}$$

with similar expression for the relaxation times of the other axes.

8.22   The measurement of rotational frictional coefficients

In this section we will describe briefly the principles behind several methods of estimating rotational relaxation times or rotational diffusion coefficients.

First, we may recollect from section 8.2 that the intrinsic viscosity of a suspension of asymmetric particles depends on the shear rate. This is because a rod-like particle (for instance) tends to align with its long axis perpendicular to the velocity gradient. This tendency to alignment is opposed by rotational diffusion so that measurements of the intrinsic viscosity in a Couette viscometer at different shear rates may be used to obtain information about the rotational diffusion coefficients.

A similar principle lies behind the method of *flow birefringence*. This method depends upon the fact that a solution of asymmetric molecules in which the latter are strongly orientated is not isotropic and exhibits birefringence. Thus measurement of the birefringency of the solution in the annular space in a Couette viscometer is capable of providing information concerning the rotational diffusion coefficient.

In these first two methods, the orientating force is provided by the differential hydrodynamic forces acting on the particle as illustrated in figure 8.2. If the macromolecule possesses an electric dipole moment it will tend to orientate in an electric field. Thus if a strong field is applied to a solution of such molecules they will orientate and the solution will become birefringent. If the field is then suddenly removed, rotational relaxation will occur and the birefringence will decay exponentially. Measurements of this decay can then be used to provide estimates of the rotational relaxation times. This is the method of *electric birefringence*.

If an alternating electric field is applied, polar macromolecules will tend to rotate backwards and forwards so as to maintain the directions of their dipoles aligned with the direction of the field. In so far as they are able to do this, the orientation of the dipoles in the field contributes to the dielectric constant of the solution, which can be measured. If, however, the frequency of the alternating field is too great, the angular velocity at which the field is able to make them rotate is insufficiently great for them to keep in step with the field so that the degree of orientation that is produced at each cycle of the field is reduced. This reduction in the orientation results in a decrease in the dielectric constant. Thus measurements of the dielectric constant as a function of the frequency of the field can provide estimates of the rotational relaxation times. This is the method of *dielectric dispersion.*

These methods depend on the application of a mechanical or an electrical force to orientate the macromolecules. There would appear to be no other forces available to affect this orientation, but there is one final method of measuring rotational relaxation times that does not require the application of an orientating force. This is the method of *depolarisation of fluorescence*. This method requires that the macromolecules carry fluorescent groups either naturally or by having them fixed thereto artificially. If a solution of fluorescent molecules is illuminated by a beam of plane-polarised light (of such a wavelength that it is absorbed by the fluorescent chromophores), the light will be preferentially absorbed by those molecules which, at any instant of time, are aligned with their excitation transition moments parallel to the plane of polarisation. These molecules will then proceed to emit the absorbed energy as fluorescence in a random manner such that the intensity of the fluorescence decays exponentially with a time constant termed the mean *lifetime of the excited state*. During this random delay between absorption and reemission, the molecules will have rotated in a random manner governed by their rotational relaxation times. This means that the directions of the transition moments for the emission become disorientated and hence that the radiation emitted becomes progressively less polarised as time increases. This describes the polarisation of the fluorescence originating from a short flash of exciting light. If the solution is illuminated by a continuous beam of exciting light, the fluorescence will appear as a continuous beam. This continuous fluorescent beam will be made up of quanta which were originally absorbed at various previous times and the earlier they were absorbed, the less polarised they will be. We thus see that the mean degree of polarisation of the fluorescent beam is less than unity. Measurements of this degree of polarisation can then provide information concerning the ratio of the mean lifetime of the excited state to the rotational relaxation times of the macromolecule. It turns out on closer analysis that the incident beam need not be polarised: the fluorescent beam will still be partially polarised and the same information can be derived from it.

8.23   Combinations of hydrodynamic properties

We have seen in previous sections of this chapter that the frictional coefficients and measurable hydrodynamic properties of a series of homologous macromolecules, differing only in their molecular weight, vary with the molecular weight in a manner that can often be expressed as a proportionality to M^γ. The value of the exponent for any particular parameter depends on the conformational characteristics of the macromolecules. For convenience we have listed the theoretical values of the exponents for various parameters and various types of molecular conformation in table 8.2. Estimates of the values of one or other of these exponents can be used to ascertain the likely conformation of the members of such a homologous series of macromolecules. For instance, a plot of ln s versus ln M is expected to give a straight line plot with a slope equal to 0.5 for random-coil polymers at the theta temperature and 0.7 if they are globular spheres.

Table 8.2. *Exponents of the molecular weight in expressions for various properties of a polymer in various states*

	$\langle r^2 \rangle$	$\langle s^2 \rangle^{1/2}$	f	s	D	$[\eta]$
Compact globular spheres	—	0.33	0.33	0.67	-0.33	0
Flexible coils in theta solvent	1.0	0.5	0.5	0.5	-0.5	0.5
Flexible coils in good solvent	1.2	0.6	0.6	0.4	-0.6	0.8
Rod-like molecules	2.0	1.0	0.48	0.52	-0.48	1.8

For a series of homologous macromolecules differing only in their molecular weight, M, various conformational and hydrodynamic parameters may vary with M so as to be proportional to M^γ. The value of the exponent, γ, may vary with the nature of the conformation. The table gives γ for various properties for various types of macromolecular conformation as suggested by the theories outlined in the text.

Further interesting information can be obtained by combining two or more hydrodynamic parameters for the same macromolecule. For instance, we have seen that the translational frictional coefficient and the intrinsic viscosity both depend on the volume of the hydrodynamic particle in the case of those molecules that may be approximated by ellipsoids of revolution. This hydrodynamic volume is generally unknown. However the volume, v, may be eliminated between equations (8.4) and (8.6) with the expression for the frictional coefficient of a sphere given in table 8.1. There thus results the Scheraga–Mandelkern equation:

$$\beta' = \frac{N_A s [\eta]^{1/3} \eta}{M^{2/3}(1 - \bar{v}\rho)} = \frac{v^{1/3}}{f_r} \left(\frac{N_A}{162\pi^2}\right)^{1/3} = \frac{v^{1/3}}{f_r} 7.22 \times 10^6 \qquad (8.61)$$

The parameter β' is defined in terms of the experimentally measurable quantities s, $[\eta]$, M, $\bar{v}$, ρ, η and Avogadro's number, N_A. It is seen to be equal to the constant 7.22×10^6 multiplied by $v^{1/3}/f_r$. This latter factor depends only on the axial ratio of the ellipsoid. Thus equation (8.61) in principle enables an unambiguous estimate of the axial ratio. This axial ratio would then be that of the ellipsoid that behaves in a hydrodynamic sense exactly as the real particles whose experimentally determined parameters have gone into the calculation of β'. That is to say it is the axial ratio of the hydrodynamically equivalent ellipsoid.

The major defect of this approach is that β' is disappointingly insensitive to p, the axial ratio. For spheres, β' takes it minimum value of 9.80×10^6. It lies below 9.94×10^6 for all oblate ellipsoids and below 11.74×10^6 for prolate ellipsoids of axial ratio above 0.067. It is only for long rod-like particles that β' increases significantly beyond the value appropriate for a sphere. It is also interesting to note that the experimental value of β' for some globular proteins is less than the minimum theoretical value. This suggests that the model of an ellipsoid of revolution is fundamentally inappropriate for such proteins.

Equation (8.61) may be written in terms of the diffusion coefficient rather than the sedimentation coefficient by substituting for s from equation (8.46), but no gain in usefulness is achieved, since diffusion coefficients are more laborious to measure than sedimentation coefficients. On the other hand, expressions analogous to equation (8.61) can be obtained by combining either the sedimentation coefficient or the intrinsic viscosity with rotational diffusion coefficients. The resulting analogues of β' vary more rapidly with the axial ratio, but have not been extensively utilised on account of the difficulty of measuring rotational diffusion coefficients with precision.

If an expression for β' is derived from equations (8.11) and (8.13) for non-draining random coils, whose hydrodynamic properties depend on the mean square radius of gyration (rather than on the volume of the hydrodynamic particle), and the radius of gyration is eliminated, β' is seen to be a constant equal to 12.8×10^6.

Thus a value of β' in the range $(11.3 \pm 1.5) \times 10^6$ is seen to encompass the values appropriate for random coils, all oblate ellipsoids of revolution and prolate ellipsoids of axial ratio greater than 0.025. Thus, for macromolecules known to be consistent with these conformations, an estimate of the molecular weight, correct to within 13%, can be obtained by assuming a value of β' equal to 11.3×10^6 and employing equation (8.61). Such estimates are referred to as sedimentation-viscosity molecular weights. It is noteworthy that β' for a variety of random-coil polymers has been found to lie close to 11.3×10^6, a little below the theoretical value. This value may be used to obtain a sedimentation-viscosity molecular weight of known random-coil polymers.

Problems

8.1 Purple of Cassius consists of a colloidal suspension of spherical particles of gold in dilute electrolyte. Estimate the sedimentation coefficient of the particles if their radius is 10 nm. Assume that the density and viscosity of the suspension are equal to those of water and that the density of gold is 19.3 g/cm^3.

8.2 In an ultracentrifuge, what angular velocity (in rev/min) would be required to ensure that the boundary between the colloidal suspension and the solvent above it sedimented from the meniscus at a radial distance of 6 cm to the cell bottom at 7 cm in one hour? Use the data provided in problem 8.1.

8.3 Using the data in problems 8.1 and 8.2, and supposing that the colloidal particles sediment independently, estimate the mechanical force acting on each particle and its frictional coefficient.

8.4 By considering the equation of motion of an individual particle, derive an expression for the time taken for the particle to achieve a velocity equal to 90% of its terminal velocity. Express this time in terms of its sedimentation coefficient and assume that the centrifugal field is applied instantaneously, with the particle being at rest.

8.5 Calculate the translational frictional ratio of a prolate ellipsoid of revolution of axial ratio 0.1.

8.6 Transfer RNA can adopt a tertiary structure which resembles in shape a letter L; each arm can be considered to be composed of a cylinder of radius 1.0 nm. The two arms are approximately 6 nm and 8 nm long. Assuming that the molecular weight is 25 000 and the partial specific volume is 0.55 cm^3/g, estimate the sedimentation coefficient, $s^0_{20,w}$.

8.7 The sedimentation coefficient ($s^0_{20,w}$) of fibrinogen is 7.9 S and its diffusion coefficient is 2.02 Ficks. Calculate its molecular weight assuming that its partial specific volume is 0.706 cm^3/g.

8.8 Assuming that fibrinogen is a prolate ellipsoid of revolution of density 1.42 g/cm^3, estimate its intrinsic viscosity. The measured value is 27 cm^3/g. How would you account for any discrepancy between the measured and your calculated value?

8.9 The table below lists values of the sedimentation coefficient ($s^0_{20,w}$) and intrinsic viscosity of 7 samples of polyuridylic acid dissolved in 0.15 M NaCl at 25°C. Estimate the molecular weight of each sample and comment on the nature of the conformation of the polymer.

$s^0_{20,w}(S)$	$[\eta]\,(cm^3/g)$
3.47	42.0
3.75	37.1
3.84	48.2
3.99	45.5
4.61	52.4
5.53	82.5
5.63	89.8

Assume that the partial specific volume of the polymers is 0.55 cm^3/g.

8.10 Derive an expression relating the diffusion coefficients and the cross diffusion coefficients of the two solutes in a three-component solution by the use of the Onsager reciprocal relations.

Assume throughout these problems that:
The density of water is 1.0 g/cm^3 at 20°C
The viscosity of water is 0.01 $g/cm\ s$
The gas constant R is 8.314 $J/K\ mol$
Avogadro constant $N_A = 6.02 \times 10^{23}/mol$

9 Polyelectrolytes

Polyelectrolytes in solution ionise to give electrically charged *macroions* and low-molecular-weight *counterions*. If the charged groups on the macroion are negative it is a *polyanion*, while if they are positive it is a *polycation*. Polyelectrolytes that ionise to give both cationic and anionic groups are termed *polyampholytes*.

Since the solution must be electrically neutral, the total charge on the macroion must be equal and opposite to the aggregate of the charges on the counterions. A solution of a polyelectrolyte may, and generally does, also contain other low-molecular-weight electrolytes which ionise to give small ions. The small ions that carry a charge of the same sign as the macroion are termed the *coions*. We shall find it convenient to suppose that the small ions of opposite sign are identical to the counterions. We shall also suppose that the counterions and coions are both univalent. We make these suppositions for the sake of simplicity, and it must be noted that if they are not valid in practice the expressions developed below require modification.

Strong electrostatic forces operate between the various small ions themselves and between the ions and the fixed charges on the macroions. These interactions profoundly modify nearly all the properties of macromolecules that we have considered in earlier chapters. In this chapter we shall examine the effect of the charges on the macroion on its thermodynamic, conformational and transport properties. We shall also see that the existence of the electrostatic charge on the macroion gives rise to a new category of transport processes, namely *electrophoresis*, whereby the macromolecule undergoes translational movement in an electric field.

The electrostatic interactions between the macroions and the small ions is of crucial importance, and the effects that we have just mentioned depend critically on the concentration of low-molecular-weight electrolyte (which we shall refer to as salt) present in the solution. At high salt concentrations the consequences of the charges on the macroions are reduced to negligible proportions for reasons that will emerge below.

Finally we should note that of the three thermodynamic components present in a solution of salt and polyelectrolyte we shall refer to the solvent (usually water) as component 1, the polyelectrolyte as component 2 and the neutral salt as component 3.

9.1   The chemical potential and activity of ionic species

The chemical potential of a component was defined in section 3.4 as the partial molar free energy. The chemical potential of an electrolyte is defined in an identical manner. However, a similar definition of the chemical potential of an ionic species runs into the difficulty that it is not possible to vary the concentration of a single ionic species while maintaining the concentration of all others constant. If we attempted to do this, the solution would rapidly acquire a large electrostatic charge so that the energy required to vary it further would be very great. For this reason the definition of the chemical potential of an ionic species as a partial molar free energy corresponds to no realistically measurable process. This observation, however, need not preclude such a formal definition of the chemical potential of an ionic species.

If the concentration of two ionic species of opposite sign are varied simultaneously so as to maintain the neutrality of the solution, the total change in free energy then corresponds to the change brought about by a variation in the concentration of the neutral salt that is composed of the two ions. We thus see that if μ_3 is the chemical potential of the neutral salt, and μ_+ and μ_- those of the univalent cation and anion into which the salt dissociates, that $\mu_3 = \mu_+ + \mu_-$. A similar relation must hold between the standard chemical potentials so that $\mu_3^{\ominus} = \mu_+^{\ominus} + \mu_-^{\ominus}$. Following the argument of section 3.4, we may then define the (relative) activity of an ionic species, i, by

$$\mu_i = \mu_i^{\ominus} + RT \ln a_i \tag{9.1}$$

It then follows that, if a_3 is the activity of the neutral salt,

$$a_3 = a_+ a_- \tag{9.2}$$

We may also define ionic activity coefficients, f_+ and f_-, for the two ions such that $a_+ = f_+ x_+$ and $a_- = f_- x_-$, where x_+ and x_- are the mole fractions of the cation and anion and such that f_+ and f_- tend to unity at infinite dilution. It thus follows from equation (9.2) and from the fact that $x_+ = x_-$ (for a uni-uni-valent electrolyte) that

$$a_3 = x^2 f_+ f_- = (x f_{\pm})^2 \tag{9.3}$$

Here we have implicitly defined the *mean activity coefficient* of the two ions, $f_{\pm}$, as $(f_+ f_-)^{1/2}$, and replaced $x_+ = x_-$ by x. It should be noted that equations (9.2) and (9.3) are only valid for symmetrical electrolytes, in which the charges on the cation and anion are numerically equal. For non-symmetrical electrolytes, more complicated expressions must be invoked.

We have seen in section 3.1 that the total free energy of a solution may be written in terms of the chemical potentials of all the components according to equation (3.4) as $G = \Sigma_i \, n_i \mu_i$. It might be thought that the application of this relation to a solution containing two salts such as KNO_3 and NaCl would result in an ambiguity: are we to consider the neutral components to be KNO_3 and NaCl or, alternatively, $NaNO_3$ and KCl? Either pair gives the same set of ions in solution. In fact no ambiguity arises, since each salt contributes two ionic terms to the expression for G so that the same four terms are present irrespective of which pair of salts we in fact dissolved in the solvent. We thus see that the salts lose their identity when they are ionised in solution and we may pair them off in whatever manner we choose – as long as the pairs are electrically neutral – to form the thermodynamic components.

Equation (9.3) implies the surprising result that the activity of the neutral salt at infinite dilution is equal to the square of its concentration, x. This may be understood, however, by noting that the cratic contribution to the chemical potential of a fully dissociated uni-uni-valent electrolyte is twice that of an undissociated component. This is so because there are twice the number of molecular entities to be distributed in the solution and, hence, twice as much entropy is associated with the number of configurations in which they may be arranged.

Electrostatic interactions between the ionic species comprising a component at finite concentration may give rise to deviations from ideal behaviour and to mean activity coefficients different from unity. For instance counterions (or other ions) may bind to a macroion. In doing so, their cratic contribution to the free energy of the solution is removed, but we may interpret this as a reduction of the activity coefficient of the ions that have bound. Thus, if f_b is the fraction of the ions removed from solution by binding, the fraction remaining is $1 - f_b$, and this is a factor in the expression for the activity coefficient.

We have mentioned above that the definition of the chemical potential of an ionic species corresponds to no feasible process. For this reason it is never possible to measure ionic activities or activity coefficients. All that it is possible to do is to measure mean activities or activity coefficients of neutral components.

9.2 Donnan equilibria

An interesting effect, characteristic of polyelectrolytes, arises when a solution containing a polyelectrolyte is separated from the solvent by a semipermeable membrane as illustrated in figure 3.2. Let us suppose that compartment I contains solvent and low-molecular-weight ions M^+ and X^- which, like the solvent, may pass through the membrane. Compartment II contains macroions, P^{Z+}, where Z is the *charge number* or ratio of the charge carried by the macroion to that of the proton. It also contains solvent and the small ions M^+ and X^-.

Since both the salt (MX), component 3 and the solvent, component 1, can pass freely through the membrane, their chemical potentials in the two compartments must be the same at equilibrium, so that $\mu_1^I = \mu_1^{II}$ and $\mu_3^I = \mu_3^{II}$. It should be noted that there is no requirement that the chemical potentials of the individual ionic species be the same in the two compartments.

We may also note that, at equilibrium, the pressure in II will be higher than in I by an amount Π, the osmotic pressure difference. By the same argument as was employed in section 3.7 we may then see that

$$\Pi \overline{V}_1 = -(\mu_1 - \mu_1^I) \tag{9.4}$$

and

$$\Pi \overline{V}_3 = -(\mu_3 - \mu_3^I) \tag{9.5}$$

where μ_1 and μ_3 are the chemical potentials of component 1 and 3 at their equilibrium concentrations in compartment II, but at the same hydrostatic pressure as compartment I.

From equation (9.5) we may see that $RT \ln a_3^I = RT \ln a_3 + \Pi \overline{V}_3$, but for the range of osmotic pressures commonly encountered with solutions of polyelectrolytes and with the range of values of the partial molar volume, $\overline{V}_3$, of commonly encountered salts, the product $\Pi \overline{V}_3 / RT$ is small compared to $\ln a_3$. If we then ignore this term we find that, at equilibrium, $a_3^I = a_3$, so that from equation (9.2)

$$a_+^I a_-^I = a_+ a_- \tag{9.6}$$

and hence that

$$c_+^I c_-^I (y_\pm^I)^2 = c_+ c_- (y_\pm)^2 \tag{9.7}$$

where c_+ and c_- are the mole concentrations of the two ions and $y_\pm$ their mean activity coefficient on this scale.

We now note that both compartments are electrically neutral, so that by balancing the number of negative and positive charges we find for the two sides of the membrane that

$$c_-^I = c_+^I$$

$$c_- = c_+ + Zc_2 \tag{9.8}$$

where c_2 is the concentration of macroions in II. It is then a simple matter to show that

$$\frac{c_+^I y_\pm}{c_+ y_\pm} = \left(1 + Z \frac{c_2}{c_+}\right)^{1/2}$$

$$\frac{c_-^I y_\pm^I}{c_- y_\pm} = \left(1 - Z \frac{c_2}{c_-}\right)^{1/2} \tag{9.9}$$

For salt concentrations that are not too high we may assume that the mean activity coefficients $y_\pm^I$ and $y_\pm$ are equal and thus see that the concentrations

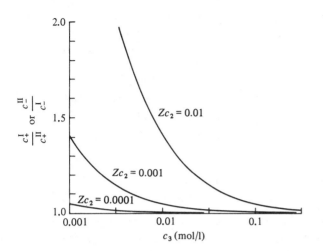

Figure 9.1. Plots of the ratios of the concentrations of a uni-valent ion on opposite sides of a semipermeable membrane in Donnan equilibrium with a macroion of valency Z and concentration c_2 in compartment II, against c_3 the concentration of salt in compartment I, for various values of Zc_2. Note that the concentration ratios are rapidly reduced to unity as the salt concentration is increased or the charge, Z, or concentration of the macroion is reduced. The ratio of the activity coefficients of an ion in the two compartments has been assumed to be unity.

of either M^+ or X^- are not the same on the two sides of the membrane. In order to balance the charge on the macroions, there are more counterions in II than in I and fewer coions. This unequal distribution of small ions is termed the *Donnan effect* after its discoverer.

If the concentration of neutral salt c_3 $(= c_+^I = c_-^I)$ is increased, c_+ and c_- also increase, and it can be seen from equations (9.9) that this leads to the concentration ratios of both ions becoming more nearly equal to unity. In other words, the Donnan effect is reduced by the addition of neutral salt to the system. The ratios also tend to unity as Z or c_2 tend to zero. Figure 9.1 illustrates the manner in which $c_+^I/c_+ = c_-/c_-^I$ varies with c_3 for several values of Zc_2.

More complicated expressions result if the neutral salt is not a symmetrical electrolyte or if several different sorts of small ion are present. It may be shown, however, that, for any ionic species, the ratio of its activities on the two sides of the membrane is given by

$$(a_i^I/a_i)^{z_i} = \text{constant} \tag{9.10}$$

where z_i is the charge number of ion i (as such it is positive for cations and negative for anions). In aqueous solution, H^+ and OH^- ions are invariably present on account of the ionisation of water. Equation (9.10) then implies that the pH is different in the two compartments. The constant in equation (9.10) tends to unity as Zc_2 tends to zero or if the concentration of electrolyte increases; under these circumstances, the pH difference tends to zero.

We have noted in the previous section that the activity coefficient of an ionic species is reduced if it binds to (or otherwise strongly interacts with) the macroion. If this occurs, the ratio $y_\pm^{\mathrm{I}}/y_\pm$ may become significantly greater than unity. This, according to equations (9.9), will tend to counterbalance the Donnan effect. An equivalent way of viewing the same effect is to note that the binding of counterions to the macroion results in a diminution of its effective charge.

9.3   Osmotic pressure

In the previous section we have seen that the osmotic pressure difference across the membrane is given by equation (9.4) in terms of the difference between the chemical potentials of the solvent in compartment I and in II. This equation may be rewritten in terms of the activities of the solvent to give

$$\Pi = \frac{RT}{\overline{V}_1} \ln (a_1^{\mathrm{I}}/a_1) \tag{9.11}$$

If we now assume that the activity coefficients of the solvent in the two compartments are both unity, the ratio of activities, a_1^{I}/a_1, is equal to the ratio of mole fractions, x_1^{I}/x_1. This latter ratio may be expressed in terms of the mole concentrations of the various species, ionic or otherwise, that are present so that

$$\frac{x_1^{\mathrm{I}}}{x_1} = \frac{c_1^{\mathrm{I}}(c_1 + c_+ + c_- + c_2)}{c_1(c_1^{\mathrm{I}} + c_+^{\mathrm{I}} + c_-^{\mathrm{I}})} \tag{9.12}$$

We then find, by noting that $\ln (1 + x) = x - x^2/2 + \cdots$ and that c_1^{I} and c_1, the concentrations of the solvent, are far greater than the concentrations of the other species as long as the solutions are dilute with respect to both polyelectrolyte and neutral salt, that to a good approximation

$$\frac{\Pi \overline{V}_1}{RT} = \ln (x_1^{\mathrm{I}}/x_1) = \frac{c_+ + c_- + c_2}{c_1} - \frac{c_+^{\mathrm{I}} + c_-^{\mathrm{I}}}{c_1^{\mathrm{I}}} \tag{9.13}$$

For dilute solutions, c_1^{I} and c_1 are both nearly equal to the reciprocal molar volume of the pure solvent, $1/V_1^*$, so that in the limit of infinite dilution we have

$$\Pi = RT(c_+ - c_+^{\mathrm{I}} + c_- - c_-^{\mathrm{I}} + c_2) \tag{9.14}$$

It is important to note that the various concentrations appearing in equation (9.14) are those appertaining to equilibrium. This being so, we may invoke equations (9.9). If we assume that $y_{\pm}^{\mathrm{I}} = y_{\pm}$, and expand the expressions under the square root sign by means of the binomial theorem to the first three terms, we obtain

$$\Pi/\rho_2 = RT\left(\frac{1}{M_2} + \frac{Z^2\rho_2}{4M_2^2 c_3} + \cdots\right) \tag{9.15}$$

where we have replaced c_2, the mole concentration of macroion, which is equal to the mole concentration of polyelectrolyte component, by ρ_2/M_2 and $1/c_+ + 1/c_-$ by $2/c_3$, which is valid for dilute solutions. ρ_2 may be taken as the weight concentration of the polyelectrolyte component if M_2 is taken as its molar mass, i.e. the molar mass of the macroion together with its full complement of counterions; but, with equal validity, ρ_2 may be taken as the weight concentration of the macroion alone as long as M_2 is taken as its molar mass (without its counterions). Other interpretations of ρ_2 and M_2 are equally valid as long as the mole concentration, c_2, of the species that cannot pass through the membrane is equal to ρ_2/M_2.

Equation (9.15) is analogous to equation (3.26) and shows that the molecular weight of the macroion can be obtained from osmotic-pressure measurements, notwithstanding the Donnan effect. The Donnan effect is manifest in the apparent second virial coefficient $Z^2/4M_2^2 c_3$, which increases as the square of the charge on the macroion and which decreases if the concentration of neutral salt present is increased. For this reason it is desirable to have neutral salts present in osmotic-pressure determinations of the molecular weights of polyelectrolytes in order to reduce the Donnan effect and to reduce the errors introduced by the extrapolation of Π/ρ_2 to infinite dilution.

In deriving equation (9.15) we have tacitly ignored interactions between the macroions, such as excluded-volume effects as discussed in sections 3.16 and 3.17, as well as electrostatic interactions between the macroions. A more detailed treatment must take these into account. For rigid molecules, such as globular proteins or DNA, the contribution to the second virial coefficient from the Donnan effect is usually much greater than the excluded volume effect except at low charge, Z, or high salt concentrations. For flexible macroions the situation is complicated by the fact that the size of the domain of the macroion varies with the salt concentration as we shall see in section 9.12. But in all cases the value of Π/ρ_2 extrapolated to infinite dilution is equal to RT/M_2.

A further complication arises if ions, the counterions for instance, bind to the macroion. This as we have seen may be interpreted in terms of a reduction of the effective charge on the macroion or as a decrease in the activity coefficient of the ion. In either case it results in a diminution of the apparent second virial coefficient. It is noteworthy that the second virial coefficient commonly observed for highly charged polyions such as nucleic acid are considerably less than the values expected from equation (9.15). These observations can often be interpreted in terms of the phenomenon of counterion condensation that we discuss below in section 9.10.

9.4   Equilibrium dialysis

Two solutions, separated by a semipermeable membrane, with one of them containing macroions, is, we have seen, in equilibrium when the chemical potential of any salts that are present and which can pass through the membrane is the same in both compartments. The properties of the solution containing the macroion clearly depend on the concentration of the latter. In later sections of this chapter we shall see that the relations between the density and refractive index of the solution and its concentration are of great importance when we come to consider the scattering of light and the transport properties of solutions of macromolecules.

The process whereby a solution of a macromolecule is brought into equilibrium with a solvent containing a diffusible third component is termed *equilibrium dialysis*. Experimentally the solution containing the macromolecule may be contained in a small sack made of semipermeable membrane or in a small vessel separated from the 'outer' solution by such a membrane. The solution containing the macromolecule is sometimes termed the inner solution. If the outer solution is of large volume and is changed several times it is possible to bring the inner solution to equilibrium with an outer solution of any specified composition. At equilibrium, we emphasise, the chemical potential of the low-molecular-weight components are the same in the inner and outer solutions.

It is possible to prepare in this manner a series of solutions containing the macroion at different concentrations, ρ_2, and each in equilibrium with a defined outer solution. For dilute solutions it is generally found that the density and refractive index of these inner solutions are linearly dependent on the concentration, ρ_2, though this is not necessarily the case if there are strong interactions between macroion and one of the dialysable components. In so far as such linear dependencies are found, it is clear that $(\partial \rho / \partial \rho_2)_{T,P,\mu_3}$ and $(\partial n / \partial \rho_2)_{T,P,\mu_3}$ are constant.

If the macromolecule were uncharged and no third component were present the rate of change of density with concentration would, as indicated in section 3.3, be equal to $1 - \rho_0 \bar{v}_2$, where $\bar{v}_2$ is the partial specific volume of the macromolecule. This quantity, as we have seen in the previous chapter, plays an important role in the description of the behaviour of macromolecules in the ultracentrifuge at infinite dilution when $\rho = \rho_0$. In the case of polyelectrolytes the quantity $(\partial \rho / \partial \rho_2)_{T,P,\mu_3}$ plays an analogous role. We specify that μ_3 is kept constant (as well as T and P) and note that, in so far as $\partial \rho / \partial \rho_2$ is independent of ρ_2, it is equal to $\Delta \rho / \rho_2$, where $\Delta \rho$ is the difference between the density of the inner solution and that of the outer solution. It should be clearly recognised that this ratio $\Delta\rho/\rho_2$ is not equal to $1 - \rho_0 \bar{v}_2$, where $\bar{v}_2$ is the true partial specific volume of the polyelectrolyte because of the Donnan redistribution of small ions and also because of the possible binding of the third component to the macroion.

In the same way, $(\partial n/\partial\rho_2)_{T,P,\mu_3}$, the rate of change of refractive index of the inner solutions with macroion concentration, is not in general equal to the refractive index increment $(\partial n/\partial\rho_2)_{T,P,\rho_3}$ of the polyelectrolyte component, for this is defined as the rate of change of refractive index of a solution in which the only change is in the concentration of the component of concentration, ρ_2. It is the quantity $(\partial n/\partial\rho_2)_{T,P,\mu_3}$ which, in the case of polyelectrolyte solutions, must be substituted for $dn/d\rho_2$ in the description of the scattering of light as discussed in chapter 7.

We have already seen that osmotic pressure of a solution of macroions can be used to derive a value for the molecular weight of the macroion when the two solutions are in Donnan equilibrium. That is to say the system behaves as if it were a two-component system as long as it is recognised that the 'solvent' (which contains the third component) is taken as the solution in which the chemical potential of the third component is the same as it is in the solution of the macromolecule. This appears to be a general result which is applicable to measurements of molecular weights by light scattering or by the method of sedimentation equilibrium.

9.5 Membrane potentials

We have seen that, at equilibrium, the concentrations and activities of a small ion are not the same on the two sides of the membrane. Suppose now that we were to place two electrodes that are reversible to either ion (M^+ or X^-) in the two compartments. Since the activities of the ions are different on the two sides, we might naively expect there to be a potential difference between them; this potential difference would be $-(RT/F) \ln (a_+^I/a_+)$. Further reflection will inform us, however, that this potential must in fact be zero for, if it were not, electrical work could be obtained by allowing a current to flow so that the free energy of the system would decrease. Since we have assumed that the system is at thermodynamic equilibrium, the free energy must be at a minimum so that no work could in fact be obtained. This implies that there must be an equal and opposite potential difference elsewhere in the system. Since potential differences between phases only exist at the interfaces between them, this equal and opposite potential difference must reside at the membrane. We therefore see that there is a potential difference equal to $(RT/F) \ln (a_+^I/a_+)$ (or $-(RT/F) \ln (a_-^I/a_-)$, which is the same in virtue of equation (9.6) across the membrane. This is the *membrane potential*.

The existence of this potential difference implies that there is a slight excess of charge in one of the compartments, but this is so small that the assumption of electrical neutrality as expressed by equation (9.8) is not compromised.

We have seen in section 9.2 that there is no requirement that the chemical potential of an ion be the same on the two sides of the membrane at equilibrium. It may be readily seen, however, that what is constant on the two sides of the membrane for any ionic species is the *electrochemical potential*, μ_i^F, which we define by

$$\mu_i^F = \mu_i + z_i q N_A \Phi \tag{9.16}$$

where Φ is the electrical potential of the phase in which the chemical potential is μ_i, z_i is the valency and q is the charge on a proton. The constancy of the electrochemical potential results from the requirement that the transfer of a small quantity of the species across the membrane results in no net energy change; this energy change is compounded of the chemical free energy contingent on transporting the ions from one phase to another and also the electrical work done in moving them against the potential difference between the two phases.

9.6   Charge fluctuations

In the previous section, the charge, Z, on the macroion has been assumed to be a fixed and definite quantity. This is reasonable provided the groups on the polyelectrolyte molecule that ionise to give the charge have the character of a strong electrolyte. If, however, they are only partially ionised, as they will be at a pH in the vicinity of the pK of the groups, not all of the macroions will carry the same charge at any instant. The value of Z to be taken in equation (9.15) should then be the average charge $\langle Z \rangle$.

For polyampholytes, such as proteins, there is a pH such that the average net charge on the macromolecule is zero. This is the *isoelectric point* of the polyampholyte. Although the average net charge is zero there will be some molecules carrying positive charges and some carrying negative charges at any instant and the mean square charge $\langle Z^2 \rangle$ is not, in general, equal to zero. Molecules that at any instant carry a charge will interact and give rise to deviations from ideality. We shall see below that the free energy of interaction between ions is proportional to the square root of the ionic strength, which, in the case of a polyampholyte at its isoelectric point in the absence of added salt, is proportional to the concentration of the macromolecule. It may be shown that a consequence of this is that the virial expansion of the relative chemical potential of the solvent contains a term $\frac{1}{2} \{ \pi N_A [\langle Z^2 \rangle q^2]^3 \rho_2 / M_2 [\epsilon_r \epsilon_0 kT]^3 \}^{1/2}$, and which results solely from these fluctuations in the instantaneous charge. Terms in $\rho_2^{1/2}$ are also present in the virial expansion, even if $\langle Z \rangle$ is not equal to zero, but are then small compared with the term due to the Donnan effect. It should also be noted that the terms in $\rho_2^{1/2}$ decrease to negligible values on the addition of neutral salt.

Finally it turns out that the term in $\rho_2^{1/2}$ is negative. This is a consequence of the fact that the interactions between the macromolecules carrying instantaneous charges is attractive and opposed to excluded-volume effects, which are repulsive and give rise to a positive second virial coefficient.

9.7   The ion atmosphere

We must now consider the electrostatic interaction between the various ions present in the solution. In particular we shall consider the interaction between the macroion and the small ions and also the interactions between the small ions themselves. We shall not consider interactions between different macroions; that is to say we shall consider the solution to be sufficiently dilute that the macroions are rarely close enough to interact; any results concerning the macroion that we derive will only hold in these limiting circumstances.

First we note that the coulombic attraction between the macroion and the counterions results in a tendency for the latter to congregate in the vicinity of the macroions. As the counterions approach the macroion (at constant temperature) the energy of the solution decreases. In the absence of other factors, the counterions would approach the macroion as closely as they were able, within the limitations of the exclusion forces operating between them. However, the congregation of the counterions in the vicinity of the macroion is also accompanied by a loss of entropy, and it is this that balances the tendency of the counterions to bind to the macroion. A compromise is reached in which the overall free energy is at a minimum and in which there is an increase in the concentration of counterions – and, by the same argument, a decrease in the concentration of coions – in the vicinity of the macroion.

Thus the macroion is surrounded by a region in which there is a net charge which is opposite to its own charge. This surrounding region constitutes the *ion atmosphere* of the macroion. It is important to note that the excess counterions in the ion atmosphere are not rigidly bound to the macroion, but are unlocalised and free to move about. We may note that these considerations apply not only to macroions but to every ion, large or small, that is present in the solution. Even the counterions and coions have ion atmospheres.

We shall illustrate the details of the formation of the ion atmosphere with reference to a spherical particle of radius, R, which carries a charge smeared uniformly over its surface. We suppose that this charge is Zq, where q is the charge on a proton.

In the absence of any small ions, the electrical potential at a distance, r, from the centre of the sphere is equal to $Zq/4\pi\epsilon_r\epsilon_0 r$; that is to say the work required to bring a unit positive charge from infinity to this point is given by that expression. In the presence of finite concentrations of small ions, the value of the potential is altered to some value, Φ, which remains to be investigated. To do this we invoke the Debye–Hückel theory.

We first note that the potential in the vicinity of our charged sphere is spherically symmetric and depends only on r, the radial distance from its centre.

For such spherically symmetric potentials there is a relation derived from the Poisson equation in electrostatics that relates the second differential of the potential at any point to the charge density, ρ, at that point:

$$\frac{1}{r}\frac{d^2(r\Phi)}{dr^2} = -\frac{\rho}{\epsilon_r\epsilon_0} \tag{9.17}$$

To be precise, the potential in the vicinity of the macroion fluctuates with time as the counterions undergo their random Brownian motion; we shall assume that equation (9.17) remains valid when Φ refers to the time-average potential and ρ to the time-average charge density.

Let us now consider the free energy change in bringing one counterion from a great distance to a point at a distance, r, from the centre of the charged sphere. If the activity of the counterion at this point is a_- and at the great distance a_0, the change in chemical free energy (including the cratic contribution) is just $kT \ln(a_-/a_0)$. To obtain the total free energy change we must add to this the electrical work required, which is equal to $-q\Phi$ (assuming that the counterion carries a charge of $-q$). At equilibrium this total change in the free energy must be zero. This proposition may be stated in another way by saying that the electrochemical potential of each ion must be the same in every region of the solution when at equilibrium.

If we now assume that the activity coefficient of the counterion does not change when it is introduced into the ion atmosphere we easily see that

$$c_- = c_0\, e^{+q\Phi/kT} \tag{9.18}$$

where c_0 is the concentration of counterions at a great distance. A precisely similar argument establishes that the concentration of coions at a point where the potential is Φ is given by

$$c_+ = c_0\, e^{-q\Phi/kT} \tag{9.19}$$

Before proceeding we may note that equations (9.18) and (9.19) could as well have been derived by invoking the Boltzmann distribution law by supposing that the probability of finding an ion at a point where the potential is Φ is proportional to $e^{q\Phi/kT}$. We should also note that however we derive these equations, they are founded on the assumption that the activity coefficient of the ion is constant. To eliminate this assumption requires a much more sophisticated statistical mechanical argument which would be beyond the scope of this book.

The charge density at the point where the potential is Φ is clearly seen to be $N_A q(c_+ - c_-)$. It then follows from equations (9.17), (9.18) and (9.19) that

$$\frac{1}{r}\frac{d^2(r\Phi)}{dr^2} = \frac{N_A q c_3}{\epsilon_r \epsilon_0}(e^{q\Phi/kT} - e^{-q\Phi/kT}) \tag{9.20}$$

Equation (9.20) is a differential equation whose solution, subject to appropriate boundary conditions, would give the potential as a function of the radial distance, r. Unfortunately it cannot be solved as it stands, and recourse must be made to a further approximation by assuming that $q\Phi/kT \ll 1$ so that the exponential terms may be expanded. If this is done it may be seen that the first and all even powers of the argument, $q\Phi/kT$, cancel to leave

$$\frac{d^2(r\Phi)}{dr^2} = r\kappa^2\phi \tag{9.21}$$

where κ, the Debye–Hückel parameter, is defined by

$$\kappa = \left(\frac{2q^2 N_A}{\epsilon_r \epsilon_0 kT}\right)^{1/2} c_3^{1/2} \tag{9.22}$$

In more extended treatments, which apply to situations in which there are several sorts of small ion with different valencies present, c_3, the mole concentration of the neutral uni-uni-valent salt, must be replaced by the *ionic strength*, I, of the solution defined by

$$I = \frac{1}{2}\sum_i c_i z_i^2 \tag{9.23}$$

where z_i is the valency of the ith ion present and c_i is its mole concentration. It should be noted that the concentration of the macroion itself should not be included in the calculation of the ionic strength, which is of significance only at infinite dilution of the polyelectrolyte.

κ has dimensions of (L^{-1}) and for water at $25°C$ is equal to $1.039 \times 10^8 \, I^{1/2}/m$. For a 0.1 M solution of salt ($c_3 = 10^3$ mole/m^3), $1/\kappa = 0.96$ nm.

Equation (9.21) is easily soluble. To do this we must apply appropriate boundary conditions and note that, for electrical neutrality, $\int_R^\infty 4\pi r^2 \rho \, dr = Zq$, while $\Phi = 0$ at infinite r. We also suppose that the radii of the small ions are equal to a so that the centre of a small ion cannot approach the centre of the macroion closer than a distance $R + a = b$. We then see that

$$\text{for} \quad r < R \qquad \Phi = \frac{Zq}{4\pi\epsilon_r\epsilon_0 R}\left(1 - \frac{\kappa R}{1 + \kappa b}\right)$$

$$R < r < b \qquad \Phi = \frac{Zq}{4\pi\epsilon_r\epsilon_0 r}\left(1 - \frac{\kappa r}{1 + \kappa b}\right)$$

$$b < r \qquad \Phi = \frac{Zq}{4\pi\epsilon_r\epsilon_0}\frac{e^{\kappa b}}{1 + \kappa b}\frac{e^{-\kappa r}}{r} \qquad (9.24)$$

We may note that, in the absence of added salt, $c_3 = 0$. Under these circumstances, and in the limit of infinite dilution of the spherical particles, $\kappa = 0$, and the potential inside our single isolated sphere and at its surface is equal to $Zq/4\pi\epsilon_r\epsilon_0 R$; similarly for $r > b$, it is given by $Zq/4\pi\epsilon_r\epsilon_0 r$, which is the value appropriate for an isolated sphere in the absence of small ions. We thus see that the effect of the small ions in forming the ion atmosphere is to cause the potential at any radial distance to decrease or to screen the charge on the sphere. This is illustrated in figure 9.2.

It must be emphasised that the validity of equations (9.24) requires that $q\Phi/kT$ be everywhere small compared with unity. The maximum potential is reached at the surface of the sphere in the absence of small ions; under these circumstances $q\Phi/kT$ is small in water at 300 K provided the ratio R/Z is large compared to 0.7 nm. For small or high charged spheres, the approximation may not be valid, but under these circumstances the counterions may associate with the macroion on account of the strong electrostatic forces acting between them so that the effective charge Zq is reduced.

We may note that the total integrated net charge in the ion atmosphere is equal and opposite to the charge on the macroion. For this reason the two sets of charges are sometimes said to constitute an *electrical double layer*. The whole assembly of charges resembles a charged electrical condenser. This concept, though graphic, is only meaningful in any useful sense when the macroion presents a more or less well-defined surface carrying a charge.

A point of great importance may be gleaned from figure 9.2, where it may be seen that, in the presence of high salt concentrations, the potential falls almost to zero at a short distance from the surface of the macroion. At greater distances, where the potential is essentially zero and the macroion surrounded by its ion atmosphere, it appears to be uncharged. That is to say, the ion atmosphere screens the charge on the macroion in such a way that the latter may behave as though it were uncharged.

The model which we have been discussing is a reasonable approximation to a micelle composed of ionic detergent molecules and to other approximately spherical colloidal particles. It has also been used to represent globular protein molecules at pH values far from the isoelectric point. For proteins close

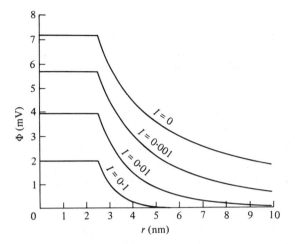

Figure 9.2. Plots of the potential in the vicinity of a spherical macroion of radius 2.5 nm and carrying a single protonic charge uniformly smeared over its surface against the radial distance from its centre, r, for various ionic strengths. The ionic strengths are given in units of mol/l. The potentials are given in millivolts; if the macroion carries a charge of Z times the charge on a proton the potential must be multiplied by this factor.

to their isoelectric point, when Z is small, the assumption that the charge is smeared uniformly over the surface is unreasonable and it becomes necessary to consider the discrete nature of the charge provided by the ionised amino and carboxyl groups. It must also be taken into consideration that the protein molecule may exist in a large number of different ionic configurations which differ according to whether the various groups are ionised or not and which are in equilibrium with each other.

For rod-like macroions such as DNA the model of a uniformly charged cylinder has been employed, and for random-coil molecules that of a spherical domain containing a uniform charge density. For such models the Poisson–Boltzmann equation (of which equation (9.20) is the version appropriate to a spherically symmetrical potential) must be solved with appropriate boundary conditions. This frequently leads to great mathematical complexity, which we shall not discuss further, save to note that the charged object is invariably shielded by its ion atmosphere in a manner similar to that described above for a sphere.

Finally, we should note that neglecting the activity coefficients required in the derivation of equations (9.18) and (9.19) is only strictly valid at infinite dilution of the salt. It is only under these circumstances that we may expect equations based on the Poisson–Boltzmann equation to be valid.

9.8   The electrical free energy

We have seen in the previous section that the charge on a macroion together with its ion atmosphere resemble the charge on an electrical condenser. We may therefore inquire as to the electrical work required to charge it up.

If Φ is the potential at the location of a charge, Q, the work required to increase the charge by an amount dQ is $\Phi \, dQ$. Thus the work required to assemble the entire charge, Q, is $\int_0^Q \Phi \, dQ$.

We thus see that the work required to charge up the spherical macroion that we discussed in the previous section is found by substituting for Φ, the potential at the surface of the sphere as given by equations (9.24). The result, denoted by G_{el}, is

$$G_{el} = \int_0^{Zq} \frac{Zq}{4\pi\epsilon_r\epsilon_0 R} \left(1 - \frac{\kappa R}{1 + \kappa b}\right) d(Zq) = \frac{Z^2 q^2}{8\pi\epsilon_r\epsilon_0 R} \left(1 - \frac{\kappa R}{1 + \kappa b}\right)$$

(9.25)

The first term on the right-hand side of equation (9.25), which is independent of κ, represents the work required to charge up the sphere in the absence of its ion atmosphere. As such it is the *self energy* of the charge on the sphere. The second term, which depends on κ, represents the amount by which the first term is reduced because of the presence of the ion atmosphere. It thus represents the energy of interaction of the charged sphere with its ion atmosphere.

A particular application of equation (9.25) is to use it to calculate the activity coefficient of one of the small ions. In the derivation of equation (9.25) no restriction was placed on the magnitude of the radius, R. We may therefore take it that G_{el} in equation (9.25) represents the electrical free energy of one of the small ions if R is set equal to a so that $b = 2a$, and Z^2 is set equal to 1 (assuming that the small ion is univalent). If we further set $\kappa = 0$ we obtain the electrical free energy of the ion at infinite dilution. The difference between this and the electrical free energy at finite concentration is nothing more than the excess free energy per ion, which is $kT \ln y_{\pm}$. We thus see that

$$RT \ln y_{\pm} = -q^2 N_A \kappa a/(1 + 2\kappa a) \, 8\pi\epsilon_r\epsilon_0 \qquad (9.26)$$

The validity of this expression has been amply verified by experiment for low ionic strengths.

Expressions for the electrical free energy of other distributions of charge have also been obtained as a function of κ and the parameters describing the distribution. These expressions generally contain a factor equal to $Z^2 q^2$, where Zq is the total charge of the assembly. A point that should be borne in mind is that the potentials in the vicinity of the various charged groups on a non-symmetrical macroion are not necessarily the same. Thus the integrals $\int \Phi \, dQ$ appertaining to each charged group must be evaluated separately and summed to obtain the total energy.

9.9   The charge on the macroion

So far we have tacitly assumed that the charge, Zq, of the macroion is a well-defined quantity and that the ionisable groups on the polyelectrolyte behave like a strong electrolyte. For many polyelectrolytes this is a reasonable assumption, and frequently one ionisable group is contributed by each monomer unit. If this is the case, the charge, number, Z, is numerically equal to the degree of polymerisation, σ.

Several factors serve to cloud this simple picture; these are associated with the binding of ions to the macroion. When this occurs a small ion binds to the macroion in a manner dictated by the law of mass action so that the extent of binding is determined by the activity of the small ion in the surrounding solution. Binding of this sort is referred to as *site binding* to distinguish it from the interaction between the small ions and the macroion which occur in the formation of the ion atmosphere.

Ions which are site bound are localised; that is to say that while they are associated with the macromolecule they are not free to move about. Site bound ions serve to change the net charge of the macroion from the stoichiometric value calculated from its composition; clearly they may increase or decrease the effective value of Z according to the sign of their charge.

One particularly important type of site binding occurs if the ionisable groups on the polyelectrolyte are weakly acidic or basic groups. In this case hydrogen ions (or protons) may bind to weakly basic groups or dissociate from weakly acidic groups to an extent determined by the hydrogen ion activity of the solution or pH. For instance the $-COOH$ groups or polyacrylic acid may or may not be dissociated to $-COO^-$ groups; proteins generally contain amino acids whose side chains carry both $-COOH$ and $-NH_2$ groups. The former may ionise to give a negative charge and the latter may bind a proton to give a positive charge. The size and magnitude of the net charge clearly depends on the pH.

The binding of the protons to the conjugate base form of the ionisable groups is covalent in character and may result in changes in the electronic structure of the ionisable groups. This may be manifest in changes in, for instance, the optical absorbance spectrum of the groups. Such changes may be used to monitor the extent of protonation which may allow the calculation of the net charge on the macromolecule.

Other ions may also bind in a covalent manner. For instance Ag^+ ions will bind to the heterocyclic bases of nucleic acids; these too may modify the spectrum of the macroion. When an ion such as Co^{++} binds in this manner it is likely that its solvation is altered; this may result in a change in the spectrum of the ion itself.

In other situations the interaction between the macroion and small ions bound to it may depend only on secondary forces such as were discussed in chapter 2. An example of this is the binding of small organic ions to the protein bovine serum albumin. It is for this sort of reason that the isoelectric point of a protein (the pH at which the net charge, Z, is zero) is not necessarily the same as the isoionic point (the pH at which there is an equal number of ionised carboxyl and amino groups).

We have indicated above that the binding of site bound ions is governed by the law of mass action so that the extent of binding (more exactly, the mean number of ions bound per macromolecule) depends on the activity of the ion in the solution and the equilibrium constant for the binding reaction, K. It is shown in textbooks on chemical thermodynamics that $-RT \ln K = \Delta G^0$, where ΔG^0 is the standard free energy change when one mole of ions is bound. A complication arises when many ions can bind to a macromolecule with the effect of altering its net charge. This arises because the greater the charge on the macromolecule, the greater the amount of electrical work that must be done to bring another ion up to a binding site against the potential gradient in its vicinity. This means that the net free energy of binding decreases as more ions are bound. This in turn implies that the equilibrium constant for the binding reaction progressively decreases as the extent of binding increases. In other words, the binding of any particular type of ion occurs less readily, the more ions of that type that are bound. It is for this reason that the titration curves of polyelectrolytes carrying weak basic or acidic groups do not usually follow the familiar curves characteristic of the titration of simple acids and bases.

If the electrical potential can be expressed in a simple manner as a function of the net charge on the macroion, the form of the titration curve may be worked out, but in the case of flexible polyelectrolytes an additional complication arises. This is because the conformation of the macromolecule changes as the net charge changes. This conformational change in turn leads to changes in the potential in the vicinity of the binding sites. The matter is of some complexity and it is beyond the scope of this book to delve into it more deeply.

Another category of site binding is provided by the formation of *ion pairs*. An ion pair consists of two ions, or a small ion and a charged group on a macroion, held in close proximity at their distance of closest approach by purely electrostatic forces. No covalent interaction is involved, and the hydration and electronic properties of the participating ions are unchanged. We have alluded to this effect in section 9.7 where we suggested that it is liable to occur with small or highly charged ions. In so far as the extent of ion pair

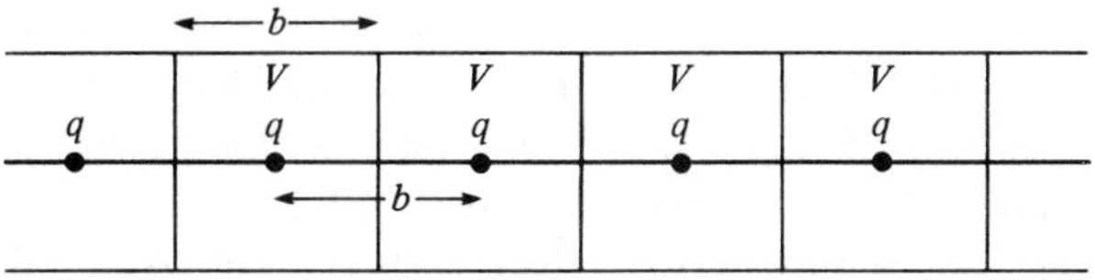

Figure 9.3. A series of fixed charges, q, are arranged in an infinite linear array so that the distance (charge spacing) between neighbours is b. Counterions condense onto this array into a cylindrical volume surrounding it. Each fixed charge is associated with a portion of this volume, V.

formation is governed by the law of mass action, it is more likely to occur at high ionic strengths.

It is important to note that ions which are site bound to a macromolecule do not contribute to the osmotic pressure of the solution.

An extension of the idea of site binding, but differing in several important respects, is that of counterion condensation that we discuss in the next section.

9.10 Counterion condensation

According to Manning, *counterion condensation* is a phenomenon associated with rod-like macroions. We shall suppose that such a macroion consists of a linear array of fixed charge groups, each of charge q and each separated from its nearest neighbour in the array by a fixed spacing b as illustrated in figure 9.3. We shall suppose, in order to avoid the complication of end effects, that the array is infinite in length. We shall also suppose that there is present in the solution a concentration, c_3, of a uni-uni-valent electrolyte, but that the concentration of counterions provided by the macroion itself is small compared to c_3.

It is supposed that each of the fixed charges is associated with θ counterions, each of charge $-q$, so that the effective value of the charge at each position in the array is reduced to $q(1 - \theta)$. The following argument (which is not rigorous) is deployed to demonstrate that θ is greater than zero when the linear charge density, q/b, is greater than a certain critical value and that the condensation of the θ counterions reduces the effective charge density to this critical value. It is supposed that these condensed counterions are not localised or rigidly bound to the charges in the array. In fact, we suppose that associated with each fixed charge is a volume, V, as illustrated in figure 9.3, and that the condensed ions are free to move up and down the array, although they are contained in the cylindrical volume composed of these volumes.

We begin by supposing that each fixed charge in the array is surrounded by an ion atmosphere so that the potential due to any one of them and at a distance, r, from it is given by a screened Debye-Hückel expression and equal to $q(1 - \theta)\, e^{-\kappa r}/4\pi\epsilon_r\epsilon_0 r$. This introduces an approximation since in fact the field due to the other charges in the array will modify the distribution of ions in the ion atmospheres and hence the potential. Within the context of this approximation, the neglect of the effect of the finite size of the ions is unimportant.

When the counterions condense on to the array there is a change in the free energy of the solution. We shall be concerned with those contributions to this free energy change which depend on θ, so that the equilibrium value of θ is that which minimises this free energy change. Two components are of significance. First, there is a change in the electrical energy of the array of charges, ΔU_c. Secondly, there is a contribution $-T\Delta S_c$, where ΔS_c is the change of entropy associated with the transfer of the condensed ions from a region of solution where their concentration is c_3 to a region where it is θ/VN_A.

The total energy of interaction of any one of the partially compensated charges in the array with all the others to the right of it along the array is readily seen to be

$$U_c = \frac{q^2(1 - \theta)^2}{4\pi\epsilon_r\epsilon_0 b} \sum_{j=1}^{\infty} (e^{-\kappa j b}/j) \tag{9.27}$$

This represents the energy of interaction per fixed charge with the rest of the array; interactions with elements to the left are counted as belonging to charges on the left. Thus the expression includes all pairwise interactions in the infinite array. The summation in equation (9.27) is readily carried out to yield

$$U_c = -\frac{q^2(1 - \theta)^2}{4\pi\epsilon_r\epsilon_0 b} \ln(1 - e^{-\kappa b}) \tag{9.28}$$

The change in energy when the θ ions condense on each fixed charge is the difference between the values of this expression with $\theta = 0$ and the final value of θ. If we note that, for low ionic strengths, $1 - e^{-\kappa b}$ is approximately equal to κb, we see that

$$\Delta U_c = [(1 - \theta)^2 - 1]\, \frac{q^2}{4\pi\epsilon_r\epsilon_0 b} \ln(\kappa b) \tag{9.29}$$

To calculate the change in entropy, ΔS_s, we first note that the partial molar entropy of a component in a solution in which its mole fraction is x is given by equation (3.22) to be $-k \ln x$, provided that the activity coefficient is unity. Thus the transfer of θ ions from a region where the mole fraction is x_A to a region where it is x_B is associated with an entropy change equal to $k\theta \ln x_A - k\theta \ln x_B$ or $k\theta \ln(x_A/x_B)$. For dilute solutions the mole fraction is proportional to the concentration so that x_A/x_B is equal to the ratio of concentrations in the two regions. We thus see that the entropy change associated with the condensation of the θ counterions onto one of the fixed charges is

$$\Delta S_c = -k\theta \ln (\theta/VNc_3) \tag{9.30}$$

The total free energy change is equal to $\Delta U_c - T\Delta S_c$, which can be computed from equations (9.29) and (9.30) as a function of θ. If we differentiate the resulting expression with respect to θ, and equate the result to zero, we obtain the condition for equilibrium, which defines the equilibrium value of θ. When we carry out these operations we find that

$$2\xi(1 - \theta) \ln (\kappa b) + \ln (e\theta/VNc_3) = 0 \tag{9.31}$$

where we have implicitly defined ξ by

$$\xi = q^2/4\pi\epsilon_r\epsilon_0 kTb \tag{9.32}$$

If we now note the definition of the Debye–Hückel parameter given by equation (9.22) and for brevity write $\kappa = \lambda c_3^{1/2}$, with $\lambda = (2q^2 N_A/\epsilon_r\epsilon_0 kT)^{1/2}$, we find that

$$2\xi(1 - \theta) \ln (\lambda b) + \xi(1 - \theta) \ln (c_3) + \ln (e\theta/VN_A) = \ln (c_3) \tag{9.33}$$

We now observe that, since $-\ln (c_3)$ tends to infinity as c_3 tends to zero, equation (9.32) can only be satisfied for arbitrarily low salt concentrations if the coefficient of $\ln (c_3)$ is the same on both sides of equation (9.33). We thus see that for low salt concentrations

$$\theta = 1 - (1/\xi) \tag{9.34}$$

and hence that the effective charge on each element of the array is reduced from its nominal value of q to q/ξ. At the same time the linear charge density is reduced from its nominal value of q/b to $q/\xi b$ which is equal to the constant $4\pi\epsilon_r\epsilon_0 kT/q$, which is independent of b. In water at $25°C$ the constant ξb is equal to 7.1×10^{-10} m (0.7 nm). It should be noted that the validity of equation (9.33) depends on b being less than 7.1×10^{-10} m but this is so for almost all linear polyelectrolytes even in their most extended configuration.

If we substitute equation (9.34) back into (9.33) we obtain an expression for V:

$$V = 8\pi e b^3(\xi - 1) \tag{9.35}$$

Numerical investigation then shows that if, for the sake of argument, we assume the volume, V, to be spherical, its radius turns out to be of the order of 1 nm and hence small.

Strictly speaking equations (9.34) and (9.35) are only valid at the limit of infinite dilution of the salt, but numerical investigation shows that θ is essentially independent of c_3 up to a concentration of 0.1 M. At higher concentrations another term in the expression for the free energy change accompanying condensation becomes important; namely, the energy of interaction of each partially compensated fixed charge with its ion atmosphere. The inclusion of this term at lower salt concentrations does not materially affect the argument.

Several features of this phenomenon of counterion condensation should be noted. First, we note that for water the product $\epsilon_r T$ is almost independent of temperature. Thus the linear charge density of a rod-like macroion cannot exceed a value of $(0.71)^{-1}$ elementary protonic charges per nanometre. If the nominal charge density does exceed this value, counterions condense so as to reduce the effective charge density to this figure. This effective charge density is, for a uni-uni-valent electrolyte in water, essentially independent of temperature, ionic strength or the nature of the macroion. It is a universal constant.

Secondly, the condensation of counterions in this manner is not governed by the law of mass action so that it takes place to the same extent no matter how much we dilute the solution: the condensed counterions never escape into the solution. In other words the tendency of the condensed counterions to escape into the surrounding solution and increase their entropy is always overcome by the strong electrostatic attraction that the macroion has for them.

Thirdly, we may note that the effect applies *par excellence* to a rigid rod-like polyelectrolyte such as DNA. However, from a short-range point of view, any linear flexible polyelectrolyte will resemble a linear array of charges, for the electrostatic repulsions between neighbouring charged groups along the chain will tend to maintain them at their maximum distance apart, except perhaps at high ionic strengths when they are shielded from one another. Thus counterion condensation is expected to occur with almost all linear polyelectrolytes.

An interesting situation arises if the neutral salt present is a mixture contributing several different sorts of counterion, and in particular ions with different valencies. A similar analysis can be carried out, but the general results are of some complexity. A qualitative conclusion emerges, however, that ions of higher valency condense preferentially, and that the effective charge density may be reduced to even smaller values than indicated above for 1 : 1 electrolytes. It is probably for this reason that small concentrations of divalent cations have such a pronounced effect on the properties of polyanions such as DNA, and not that they are bound in some more specific manner.

9.11   Conformational changes in polyelectrolytes

If we consider a highly charged macroion, we may see that the various charged groups (assumed to carry charges of the same sign) will repel one another. If the conformation of the polymer can change in such a way as to increase the distances between the charged groups the electrostatic energy of the macroion would decrease. It must be recognised, however, that the magnitude of this decrease would be reduced if high concentrations of salt are present, for in this case the charged groups will be shielded as explained in section 9.7. It is for this sort of reason that globular proteins tend to denature when they are highly charged at pH values far from their isoelectric point; that double helical structures such as that of native DNA tend to revert to random coils in low-ionic-strength solutions and that the size of the domain of flexible macroions such as polyacrylic acid tends to increase as the ionic strength is lowered.

A complete treatment of the denaturation of globular proteins is complicated since it must take into account the discrete nature of the charges and their location in the native structure. It must also consider the electrical free energy of the random-coil or denatured form of the polypeptide. On account of this complexity we will not pursue the quantitative aspects of protein denaturation further and only note that globular proteins tend to denature at extreme pH values and that they can frequently be protected from denaturation by the addition of salt to increase the ionic strength.

Charge effects in helix–coil transition are of some interest, particularly in the case of polynucleotides. It turns out that they are dominated by the counterion condensation effect discussed in the previous section. The double-helical form of native DNA is well represented by a charged rod or a linear array of charges. When this is heated it reverts to two random-coil polynucleotides as we have seen in chapter 5. We shall assume, following the argument in the previous section, that the random-coil form can also be represented (for short stretches of the chain) by a charged rod. The crucial point is now that the nominal linear charge density is reduced when the helix reverts to the coil so that the values of the parameter ξ are different for the two forms. We may take the two values to be ξ_h and ξ_c for the helix and coil forms. The corresponding values for native and denatured DNA are about 4.2 and 1.8 respectively. It turns out that a parameter of interest is $1/\xi_c - 1/\xi_h$ which we shall denote by η. The value of this for DNA is 0.32.

It may readily be seen from equation (9.34) that the number of condensed counterions per charge group decreases by η when the helix reverts to a random coil. This release of condensed counterions has several consequences which entail that the difference between the standard chemical potentials of the helix and coil forms of the macroion depends on the ionic strength.

It is convenient to consider, instead of chemical potential differences, the difference, Δg, in free energy per charge group of a solution containing the coil form and one containing the helix form. This free energy difference can be considered to be the sum of an intrinsic difference which would appertain even if the macromolecule were devoid of electric charge and which we denote by Δg^0 and several other terms that arise because of the charged groups.

In the first place, the release of η counterions into the solution results in an increase in the entropy of the solution. The corresponding increase in the free energy per charged group is equal to $\eta \mu_+/N_A$ where μ_+ is the chemical potential of the counterions in the solution. If we suppose that salt is present in excess, the concentration of counterions and hence their chemical potential will not be appreciably changed when the released counterions are mixed into the bulk of the solution. The chemical potential of the counterions may be written in the form $\mu_+^\ominus + RT \ln c_3$, where c_3 is the concentration of excess electrolyte assumed to be uni-uni-valent if the activity coefficient is unity. Thus the increase in free energy per mole cell from this source is $\eta(\mu_+^\ominus/N_A + kT \ln c_3)$. It may be noted that η is positive and that $\mu_+^\ominus$ is independent of c_3. This implies that Δg increases when c_3 is increased, so that the helix is stabilised on this account by the addition of salt.

In the second place, the energy of interaction of the charged groups with their ion atmospheres changes on account of the release of condensed counterions. Manning has argued that the energy of interaction of one of the charged groups in a linear array with the ion atmosphere is $-\xi^{-1} kT \ln \kappa$. Thus the change in energy of interaction when the helix reverts to a coil is $-\eta kT \ln \kappa$. This term is negative which indicates that an increase in κ or c_3 results in a destabilisation of the helix on this account.

Finally, for completeness, and in order to account for other effects including changes in the interaction of the various ionic species with the solvent, we include a term $\Delta g'$. We suppose that this is independent of c_3.

If we replace κ by $\lambda c_3^{1/2}$ we finally obtain an expression for Δg in the form

$$\Delta g = \Delta g^0 + \Delta g' - \tfrac{1}{2}\eta kT \ln (\lambda^2 c_3) + (\mu_+^\ominus/N_A + kT \ln c_3) \qquad (9.36)$$

This may be simplified and written in the form

$$\Delta g = \Delta g^0 + kT\Upsilon + \tfrac{1}{2}\eta kT \ln c_3 \qquad (9.37)$$

where Υ includes the remaining terms that do not depend on c_3.

Now from the definition of free energy it is easily seen that $\partial(\Delta G/T)/\partial(1/T) = \Delta H$ and that $-\partial G/\partial T = \Delta S$. If we apply these formulae to equation (9.37) we find that

$$\partial(\Delta g/T)/\partial(1/T) = \Delta h = \Delta h^0 + k \, d\Upsilon/d(1/T) \qquad (9.38)$$

$$-\partial g/\partial T = \Delta s = \Delta s^0 - k \, d(\Upsilon T)/dT - \tfrac{1}{2}\eta k \ln c_3 \qquad (9.39)$$

where Δh and Δs are the enthalpy and entropy changes per charged group and Δh^0 and Δs^0 are the corresponding intrinsic changes for the hypothetical uncharged molecule. It is clearly seen that Δh is independent of c_3 and that any effect of the salt concentration on the stability of the helix is purely entropic in origin.

Δg, Δg^0 and Υ in equation (9.37) depend on the temperature. At a certain temperature, the melting temperature discussed in chapter 5, the helix and coil forms are equiprobable and $\Delta g = 0$. We thus find by setting $\Delta g = 0$ in equation (9.37), dividing through by T_m, differentiating with respect to $1/T_m$ and substituting from equation (9.38) with T set equal to T_m that

$$\frac{dT_m}{d\ln(c_3)} = \frac{kT_m^2 \eta}{2\Delta h(T_m)} \tag{9.40}$$

$\Delta h(T_m)$ is the enthalpy change per charged group at the melting temperature T_m. Calorimetric data for DNA has suggested that $kT_m^2/\Delta h(T_m)$ is almost independent of T_m, positive and equal to approximately $58°C$. Since it is positive, T_m is expected to increase with c_3 and it may be concluded that added salt stabilises the helical form of DNA.

Equation (9.40) requires that a plot of the melting temperature of a helical macroion against the logarithm of the salt concentration be linear, providing, as seems to be the case for polynucleotides, $T_m^2/\Delta h$ is independent of T_m. This has been amply verified for DNA and other two-stranded and three-stranded double-helical polynucleotides. The slope of such plots (using $\log_{10} c_3$ as ordinate rather than $\ln(c_3)$ as is conventional) is found to be about $19°C$ for DNA. The value calculated from equation (9.40) on the basis of the values of η and $T_m^2/\Delta h$ given above is $21°C$.

Finally, we remark that, in the presence of divalent cations or those with even higher valency, the helical form may be stabilised even more strongly and that that the melting temperature may become a complicated function of the relative concentrations of the different sorts of ion that are present.

9.12   The expansion of flexible polyelectrolyte molecules

We have seen in section 4.9 that repulsive interactions between the different monomer units of a flexible polymer serve to increase the mean square end-to-end distance and that this expansion may be characterised by the factor α. In flexible polyelectrolytes, the repulsive interaction originating from electrostatic forces between the charged groups may be considerable and lead to large values of α for such macromolecules. These electrostatic interactions can be reduced by the addition of salt which shields the charged groups from each other.

As the salt concentration is increased and the electrostatic repulsions reduced, there may come a time at which the repulsive interactions are balanced by any attractive secondary interactions between the monomer units. In such circumstances, which are equivalent to theta conditions, the macromolecule will behave ideally and the expansion factor α will be unity. As the salt concentration is increased still further, the effect of the charges will become less and less discernible till in the limit of infinite salt concentration we may suppose that the macromolecule will behave as if it were uncharged.

This latter point may also be seen if we suppose that the individual charges on the monomer units are each surrounded by an ion atmosphere such that the potential at a distance, r, from each charge is given by an expression based on the Debye–Hückel theory as being proportional to $e^{-\kappa r}/r$. If κ is infinite, the potential is zero at all points. It must be realised that the specification of the potential in this way, although a convenient approximation, is subject to the criticism that the ion atmosphere of one charged group is likely to be distorted by the other charged groups in its vicinity, particularly if the value of $1/\kappa$ is comparable or larger than the distance between adjacent charged groups.

In terms of this Debye–Hückel approximation, the radius of the ion atmosphere surrounding each charged group is of the order of $1/\kappa$, and this is likely to be somewhat smaller than the radius of the domain of the random coil itself. This implies that most of the ion atmosphere of each charged group and hence a net charge opposite in sign and nearly equal in magnitude to the total charge on the macroion will also be held within the domain. Thus the domain itself will be almost electrically neutral. Exceptions to this may occur with small macromolecules or at low salt concentrations, but even in the complete absence of added salt the counterions of the polyelectrolyte will not succeed in escaping entirely from the domain of the macroion except at concentration so small that experimental verification becomes impossible. The origin of this effect is clearly to be traced to the strong electrostatic attraction that the macroion, with its many charged groups, bears for the counterions.

A further consequence of the repulsion between the charged monomer units is that there is a tendency for adjacent units (for which the repulsion is strongest) to be at their maximum separation. This has the effect of causing short stretches of the chain to be rod like and less coiled than they would be in the absence of the electric charges. Thus from a global point of view a flexible macroion will resemble a random coil and occupy an approximately spherical domain, while from a local point of view short regions of the chain will appear rod like.

A problem of some interest is the quantitative relation between the expansion of the polyelectrolyte and the salt concentration. To grapple with this problem first consider the macromolecule in a hypothetical state in which it carries no electric charge but is otherwise the same. First suppose that such a molecule is immersed in a theta solvent. Its mean square end-to-end distance will be given by equation (4.16) as $\langle r^2 \rangle_0 = \sigma\beta^2$. Now suppose that the solvent is changed to the one of interest containing a 1:1 electrolyte at a concentration, c_3. In general the coil will expand so that the new mean square end-to-end distance will now be $\langle r^2 \rangle = \sigma \beta^2 \alpha^2$. We may also suppose that, in this solvent, the probability of finding the coil in a conformation with an end-to-end distance in the range r to $r + dr$ is given by $\phi(r)\, dr$. Clearly $\langle r^2 \rangle = \int_0^\infty r^2 \phi(r)\, dr$.

Let us now suppose that the macromolecule is 'frozen' in one of its possible conformations and that the charges are then attached. We suppose that the conformation does not change during this charging process. The work required will include the self energy of the assembly of charges at locations defined by the frozen conformation and also the energy of interaction of the macroion with its ion atmosphere which is formed in the course of the charging process. In general, many different conformations will correspond to any given end-to-end distance, r. Let us suppose that the average energy required in the charging process for such a set of conformations is $G_{el}(r)$.

We may now see by invoking the Boltzmann distribution that the probability of finding a charged macromolecule in a conformation with an end-to-end distance, r, is proportional to $e^{-G_{el}(r)/kT}$ and thus see that the mean square end-to-end distance of the charged macroion is given by

$$\langle r^2 \rangle_c = \alpha_c^2 \langle r^2 \rangle = \alpha_c^2 \alpha^2 \langle r^2 \rangle_0 = \int_0^\infty r^2 \phi(r)\, e^{-G_{el}(r)/kT}\, dr \qquad (9.41)$$

Here we have introduced an additional expansion factor α_c to account for that part of the total expansion that is contingent on the electrostatic effects.

We see from equation (9.41) that the derivation of an expression for α_c requires the derivation of an expression for $G_{el}(r)$. This is a matter of some difficulty, and the problem has not been solved in a completely satisfactory manner. We will content ourselves by outlining three such attempts at a solution.

The first due to Katchalsky and Lifson supposes that the potential in the vicinity of each charged group and at a distance, r, from it is $q\, e^{-\kappa r}/4\pi\epsilon_r\epsilon_0 r$. It then follows that the energy of interaction of two charged groups together with their ion atmospheres is $q^2\, e^{-\kappa r_{ij}}/4\pi\epsilon_r\epsilon_0 r_{ij}$, where r_{ij} is the distance between the two groups labelled i and j. If there are Z charged groups in all we must consider the interaction between the $\frac{1}{2}Z(Z-1)$ pairs of charges. If Z is large the total energy of all these interactions may be written as

$$G_{el}(r) = q^2 \left\langle \sum_{ij} (e^{-\kappa r_{ij}}/r_{ij}) \right\rangle \bigg/ 4\pi\epsilon_r\epsilon_0 \qquad (9.42)$$

Here the summation is taken over all $\frac{1}{2}Z(Z-1)$ pairs of charged groups and the averaging implied by the angle brackets is over all conformations consistent with an end-to-end distance, r. To proceed further Katchalsky and Lifson supposed that the distribution of end-to-end distances of the charged macromolecule was given by equation (4.23) and employed the same sort of reasoning that was used in section 8.7 to derive an expression for $\Sigma_{ij} r_{ij}^{-1}$. This assumption is the major source of error in this treatment in that it puts too much weight on the most compact conformations and overestimates $G_{el}(r)$ for small r. This in turn leads to an overestimation of the expansion due to the electrostatic repulsions.

An alternative approach is that of Rice and Harris, who supposed that the macroion behaved like a series of rods hinged together and resembling Kuhn statistical segments as described in section 4.10. They supposed that the charge, Q, associated with the monomer units in each segment could, to a reasonable approximation, be considered to reside at the centre of each segment. Unlike the segments described in section 4.10, however, the segments in the present theory are supposed to resemble rigid rods (which is perhaps reasonable for reasons discussed above) and that two adjacent rod segments may be inclined at a variable angle, ϕ, to each other. They then assumed that the averages, $\langle \cos \phi \rangle$, for each pair of adjacent segments were independent. Then, following the argument of section 4.4, they deduced that

$$\alpha_c^2 = \frac{\langle r^2 \rangle_c}{\langle r^2 \rangle} = \frac{1 + \langle \cos \phi \rangle}{1 - \langle \cos \phi \rangle} \tag{9.43}$$

The average $\langle \cos \phi \rangle$ was taken to depend on the interaction between the charges at the centre of the two adjacent segments. The distance between these two charges is $b \sin \frac{1}{2}\phi$, where $2b$ is the length of a segment, so the energy of interaction of them was then taken as $Q^2 \, e^{-\kappa b \sin \frac{1}{2}\phi}/4\pi\epsilon_r\epsilon_0 b \sin \frac{1}{2}\phi$ or $G(\phi)$. The average value of $\cos \phi$ is then given by the Boltzmann distribution to be

$$\langle \cos \phi \rangle = \frac{\displaystyle\int_0^{2\pi} \cos \phi \, e^{-G(\phi)/kT} 2\pi \sin \phi \, d\phi}{\displaystyle\int_0^{2\pi} e^{-G(\phi)/kT} 2\pi \sin \phi \, d\phi} \tag{9.44}$$

Rice and Harris suggested that the appropriate value of the dielectric constant, ϵ_r, should not be taken as that of the bulk solvent (water) but some lower value to account for the fact that most of the material between two charges was either bound water or the material of the polymer chain. A figure of 5.5 was suggested. The application of this theory to carboxymethyl cellulose gave results in reasonable agreement with experiment. This may, however, be associated with the fact that this polymer is somewhat inflexible, for the application to polymethacrylic acid, which is highly flexible, was much less successful. One of the weaknesses of the theory is that α_c turns out to be independent of the molecular weight of the macromolecule. This is at variance with experience.

A radically different type of approach to the problem was developed by Flory. We have seen in section 6.5 that an expression for the expansion factor can be derived by supposing that an uncharged polymer chain resembles a gel. As this swells, the decrease in excess free energy contingent on the mixing of solvent with the polymer segments is accompanied by an increase in free energy arising from the loss of entropy of the distorted chain. Flory first supposed that the macroion could be considered to be a similar portion of charged gel which was in Donnan equilibrium with the surrounding electrolyte. As the gel or macromolecular domain swells, there is an additional decrease in the excess free energy as the solvent is extracted from the surrounding solvent and mixed with the small ions in the domain.

This implies that an additional term must be added to the right-hand side of equation (6.19); this, using the terminology of section 9.2, is the excess part of $\mu_1 - \mu_1^I$. We may thus see from equations (9.4) and (9.15) that the extra term is $RTZ^2 \overline{V}_1 \rho_2^2 / 4 M_2^2 c_3$, where ρ_2 is the concentration of macromolecular material within the domain and thus equal to $M_2 / \alpha^3 V_e$. If we further suppose that the charge number of each monomer unit of molar mass M_0 is z, and follow the same reasoning as deployed in section 6.5, we see that

$$\alpha^5 - \alpha^3 = 2C_M M_2^{1/2} (\chi_1 - \tfrac{1}{2}) + 2C_I M_2^{1/2} z^2 / c_3 \qquad (9.45)$$

where

$$C_I = (9/8\pi)^{3/2} (1/N_A M_0^2) (M_2 / \langle r^2 \rangle_0)^{3/2} \qquad (9.46)$$

C_I, like C_M, is independent of the molecular weight of the macroion, but depends on its nature.

In this expression, α should be interpreted as the product $\alpha_c \alpha$ and it is readily shown that it is equal to unity when the second virial coefficient is zero (theta conditions). For most flexible macroions, the term in C_I is much greater than the term in C_M.

Besides the limitations inherent in the theory, which were discussed in section 6.5, the extension to polyelectrolytes is subject to the conditions that the salt concentration should not be too low. If it is, the radius of the domain may no longer be large compared to $1/\kappa$ so that the domain is not electrically neutral and it may also be that the chain is so extended that the assumption of Gaussian statistics is no longer valid.

If α is large, equation (9.45) suggests that α would be proportional to $M_2^{0.1}$ and to $c_3^{-0.2}$. Both these predictions are in good agreement with experiment for several flexible polyelectrolytes. The major defect of the theory is that it overestimates the magnitude of the expansion effect compared with experiment. It is not clear whether this is due to a basic weakness of the theory or if counterion condensation should be taken into account; this effect would reduce the effective value of z and hence of the expansion.

It may be noted finally that this theory can be readily extended to give expressions for the swelling of macroscopic gels whose polymer chains bear charged groups. Such gels swell when immersed in salt solution to an extent determined by the charge density and the salt concentration.

9.13   The scattering of light by polyelectrolytes

In section 7.5 we considered the scattering of light by solutions of macromolecules in terms of fluctuations of the polarisability of small volume elements of the solution. These volume elements were considered to be small compared with the wavelength but sufficiently large that flucutations in neighbouring elements were independent. If the latter condition is not met, interference effects between the light scattered by fluctuations in adjacent elements may occur. This gives rise to additional complications which can only be avoided if the range of the electrostatic interaction between macroions is small compared with the size of the element. A practical criterion for this is that the Debye–Huckel length $1/\kappa$ should be small compared with the size of the element. It can then be shown that errors due to this effect are less than 1%, provided $2 \sin(\frac{1}{2}\theta)/\kappa\lambda$ is less than 0.05. This condition is readily met for all angles, θ, if the salt concentration is greater than 1 mM for visible light. For the scattering of X-rays, however, this condition may be more difficult to meet.

If the condition mentioned above is met so that the fluctuations are electrically neutral, it may be shown that an equation analogous to equation (7.19) is valid for polyelectrolyte solutions provided dn/dc is set equal to $(\partial n/\partial \rho_2)_{T,P,\mu_3}$ as defined in section 9.4 and that the second virial coefficient, B, is interpreted as the value given by equation (9.15). In principle the intensity scattered by the solvent, which must be subtracted from that scattered by the solution to give the excess scattering, should be that scattered by a solution of the third component that is in Donnan equilibrium with the solution containing the macroion. If equation (7.19) is interpreted in this way it may be shown (though the proof, although straightforward, is long) that the molar mass, M_2, that is obtained is that of the component whose concentration defines ρ_2. This may be the molecular weight of the macroion alone or that of the neutral polyelectrolyte. The situation is analogous to that appertaining to the measurement of molecular weights by osmotic pressure as discussed in section 9.3. No ambiguity arises even if the macroion binds counterions or coions as long as these can pass through the semipermeable membrane used in the equilibrium dialysis experiment necessary for the measurement of $(\partial n/\partial \rho_2)_{T,P,\mu_3}$.

These considerations apply to highly charged macroions. If the macroion is close to its isoelectric point, due consideration must be paid to additional fluctuations in the polarisability which arise from charge fluctuations as discussed in section 9.5. Thus at low salt concentrations terms in $\rho_2^{1/2}$ appear in the expression for the excess chemical potential of the solvent. These may give rise to curvature of the plots of $K\rho_2/R_\theta$ versus ρ_2 plots. Nevertheless, the intercept at vanishingly small concentration is still equal to $1/M_2$. Such effects have been noted in light-scattering experiments with proteins at their isoelectric point at low salt concentrations. At higher salt concentrations the effect is negligible.

9.14 Equilibrium sedimentation of polyelectrolytes

We have seen in section 8.19 that the condition for equilibrium of a two-component system in the cell of an analytical ultracentrifuge is that the total chemical potential of either species must be invariant with radial distance. In saying this we are implicitly treating each infinitesimally thin layer of the solution as a separate phase.

If ionic species are present we may suppose that, at equilibrium, these infinitesimal phases are in Donnan equilibrium and hence that there is a potential difference between each one: the membrane potential. This implies a gradient of electrical potential in the radial direction within the cell. Another way of divining the existence of this potential gradient is to note that the equilibrium distribution of the heavy macroions is slightly displaced down the cell relative to the distribution of the counterions. This small charge separation implies a potential difference.

With this potential gradient in mind we must now suppose that the condition of equilibrium that applies to any ionic species is that the total chemical potential plus an electrical term is constant throughout the cell. Equation (8.55) then becomes for each ionic species i

$$\frac{d\mu_i}{dx} - M_i\omega^2 x + z_i q N_A \frac{d\Phi}{dx} = 0 \tag{9.47}$$

where $d\Phi/dx$ is the gradient of electrical potential and z_i is the valency of the ion in question.

For a solution containing macroions, P^{Z+}, counterions, X^-, and coions, M^+, equation (9.47) applies to all three. It is then easy to eliminate $d\Phi/dx$ to obtain

$$\frac{\partial\mu_P}{\partial x} + Z\frac{\partial\mu_-}{\partial x} - \omega^2 x(M_P + ZM_-) = 0$$

$$\frac{\partial\mu_+}{\partial x} + \frac{\partial\mu_-}{\partial x} - \omega^2 x(M_+ + M_-) = 0 \tag{9.48}$$

where M_P is the molar mass of the macroion and Z its charge number. Since $\partial\mu_+/\partial x + \partial\mu_-/\partial x = \partial\mu_3/\partial x$, the gradient of chemical potential of the neutral salt, and $\partial\mu_P/\partial x + Z\partial\mu_-/\partial x = \partial\mu_2/\partial x$, the gradient of chemical potential of the neutral polyelectrolyte component, and, since $M_+ + M_- = M_3$ and $M_p + ZM_- = M_2$, the molecular weight of the polyelectrolyte (together with its counterions), equations (9.48) reduce to

$$\partial \mu_2 / \partial x - \omega^2 x M_2 = 0$$

and

$$\partial \mu_3 / \partial x - \omega^2 x M_3 = 0 \qquad (9.49)$$

These equations are exactly analogous to equation (8.55).

We may now express the variation of the chemical potential of components 2 and 3 in terms of variations in the pressure and the concentrations of the two components by analogy with equation (8.56). If we do this and eliminate $\partial \rho_3 / \partial x$ from the resulting expression we obtain

$$\frac{1}{\omega^2 x} \frac{d\rho_2}{dx} = \frac{a_{33}(M_2 - \rho \bar{V}_2) - a_{23}(M_3 - \rho \bar{V}_3)}{a_{33} a_{22} - a_{23} a_{32}} \qquad (9.50)$$

where for brevity we have defined a_{ij} as $\partial \mu_i / \partial \rho_j$. The terms in $\bar{V}$ originate from equation (8.53) and the relation $\partial \mu_i / \partial P = \bar{V}_i$.

Now it may be shown (the proof, although straightforward, is long) that the right-hand side of equation (9.50) may be written as $\rho_2 (\partial \rho / \partial \rho_2)_{T,P,\mu_3}$ $(d\Pi / d\rho_2)$ so that with equation (3.26) in mind we find that

$$M_2 \omega^2 x \left(\frac{\partial \rho}{\partial \rho_2} \right)_{T,P,\mu_3} = \frac{RT}{\rho_2} \frac{d\rho_2}{dx} (1 + 2BM_2 \rho_2 + \cdots) \qquad (9.51)$$

This expression is analogous to equation (8.57) with $(1 - \rho \bar{v}_2)$ in the latter replaced $(\partial \rho / \partial \rho_2)_{T,P,\mu_3}$. We have argued in section 9.5 that $(1 - \bar{v}_2 \rho_0)$ is equal to $(\partial \rho / \partial \rho_2)_{T,P,\mu_3}$ for dilute solutions so that the analogy is quite close. The right-hand side of equation (9.51) is expressed in terms of the virial coefficients rather than the activity coefficients as in equation (8.55), but in either case it reduces to $(RT/\rho_2)(d\rho_2/dx)$ in the limit of infinite dilution of the macromolecular component.

Equation (9.51) applies to any position in the ultracentrifuge cell at equilibrium so that the molecular weight, M_2, may be determined from an analysis of the manner in which ρ_2 varies with x just as for a two-component solution of an uncharged macromolecule.

For solutions of macroions at low salt concentrations, the virial coefficient, B, may be large, and this may give rise to practical problems. For this reason it is usual to employ moderate concentrations of neutral salt to mitigate this effect of non-ideality.

It may also be noted that equation (9.51) is inexact for macroions close to their isoelectric point in low salt concentrations, for then charge fluctuation effects give rise to terms in $\rho_2^{1/2}$ in the expression for $d\Pi/d\rho_2$. However if

$d\Pi/d\rho_2$ is substituted back into equation (9.51) it becomes exact and valid for any component in a solution containing any number of components at any concentration.

It may also be seen that the molar mass, M_2, is that of the component whose weight concentration is employed in the measurement of $(\partial\rho/\partial\rho_2)$. This may arbitrarily be the macroion or the neutral polyelectrolyte together with its counterions. This conclusion is similar to that drawn in the case of measurements of molecular weight by other methods we have discussed.

9.15 The effect of an electric field

Let us suppose that an electric field, E, is applied to a solution of a polyelectrolyte by means of two electrodes immersed in the solution. We shall not be concerned with any effect emanating from processes taking place in the vicinity of these electrodes.

The field will exert an electrostatic force on the ions present in the solution tending to move the anions towards the anode and the cations in the opposite direction towards the cathode.

The first effect to be considered is a slight separation that is produced between each macroion and its ion atmosphere. The distortion or polarisation of each macroion and its ion atmosphere causes a dipole moment. For a small field the magnitudes of the dipoles are proportional to the field strength. Smaller dipoles are produced when the ion atmosphere is tightly bound to the macroion in the presence of added salt. The effect is manifest by solutions of polyelectrolytes having a high dielectric constant.

If the field is suddenly switched off, the macroion attracts back its atmosphere to its equilibrium position, but this process takes place at a finite rate which is characterised by a relaxation time.

If the field applied alternates in direction, the direction of the dipoles will be reversed at each cycle of the field, providing the frequency is not too great. For high frequency fields, the dipoles cannot relax and reform fast enough to keep up with the field so that the dielectric constant for high frequencies is lowered. This gives rise to a dispersion of the dielectric constant with frequency whereby the dielectric constant falls as the frequency of the field is increased through a range whose magnitude is roughly equal to the reciprocal of the relaxation time. This effect should be distinguished from the dispersion of the dielectric constant which is due to the rotation of asymmetric macromolecules, and which was discussed in section 8.21. This latter effect generally takes place over a higher frequency range than the dispersion due to the polarisation of the ion atmospheres.

The second effect of the force acting on the ions present is to cause a steady movement of the ions (superimposed on their Brownian motion). This results in a transport of electric charge; negative ions move towards the anode and positive ions towards the cathode. This transport of charge constitutes an electric current, and the process whereby the macroions are caused to move is termed *electrophoresis*. We may conceive of the macroions exchanging ion atmospheres with those abandoned by the macroions moving ahead of them.

If the magnitude of the field is not too great, each ion will be accelerated to a steady terminal velocity, u, which will be proportional to the field strength, E. The ratio of the velocity to the field strength is termed the *electrophoretic mobility* of the ion, v, so that:

$$v = u/E \qquad (9.52)$$

For many macroions, v is of the order of 10^{-5} cm^2/sV. It depends on the viscosity of the solvent and on the size, shape and charge of the ion. It also depends in a complicated way on the hydrodynamic and electrical interaction between the ion and its ion atmosphere, which moves in the opposite direction, carrying solvent with it.

Two effects of the interaction between the macroion and its ion atmosphere have been noted. The first is the *electrophoretic effect* which is hydrodynamic in nature and results from the purely hydrodynamic interaction between the macroion and the counterions in the ion atmosphere which move in the opposite direction. The second is the *relaxation effect* which is electrostatic in character. It results from the polarisation of the ion atmosphere noted above; this polarisation sets up an extra electric field which acts so as to oppose the motion of the macroion, and hence slow it down.

In the presence of added salt, most of the electric current is carried by the small ions that are present, but some is nevertheless carried by the macroions themselves. The flux, J_i, of any ionic species i is equal to $\rho_i u_i$ or $E\rho_i v_i$. Thus the flux of electric charge due to this species is $N_A q E z_i \rho_i v_i$, where z_i is charge number of the ion and where we have assumed that mobilities in the anodic direction are taken as being negative. We thus see that the total charge flux or current density (current per unit cross-sectional area) is $N_A E q \sum_i z_i c_i v_i$. The electric field may be expressed as a potential gradient, dV/dx, so that by applying Ohm's law we find that the electrolytic conductivity of the solution is $N_A q \sum_i z_i c_i v_i$.

9.16   Electrophoretic mobility

Let us first consider a charged spherical macroion similar to that discussed in section 9.7. We first suppose that there is no added salt and that the solution is very dilute so that the ion atmosphere is very diffuse and there are no counterions in the vicinity of the macroion.

The electrostatic force acting on the macroion is then ZqE. When it has reached its terminal velocity we may equate this to the viscous drag which is

$6\pi\eta Ru$. We thus see that in these conditions the electrophoretic mobility is given by

$$\nu = \frac{u}{E} = \frac{Zq}{6\pi\eta R} \tag{9.53}$$

This expression may be written in an alternative form in terms of the *zeta potential*. This is defined as the potential at the surface of shear, which is the surface at which shearing between the hydrodynamic particle and the solvent occurs. If we take this surface as being the surface of the sphere (thereby ignoring any firmly bound solvent) we see that, for our isolated sphere, the zeta potential, ζ, is, from equation (9.24), with $\kappa = 0$, equal to $Zq/4\pi\epsilon_r\epsilon_0 R$. We may thus write equation (9.53) in the form

$$\nu = 2\zeta\epsilon_r\epsilon_0/3\eta \tag{9.54}$$

This is a form of Smoluchovsky's equation which holds when $\kappa R \ll 1$.

Let us now consider the situation at high salt concentration such that $\kappa R \gg 1$. The ion atmosphere is now close to the surface of the sphere and its thickness is small compared to the radius. In a very schematic manner we may represent it as a layer of counterions equal and opposite in charge to that on the sphere and resembling a parallel-plate condenser of area, A, and separation distance, d, filled with a medium of viscosity, η. As the particle moves through the solution, the plates of this condenser slip relative to each other with a velocity, u. The viscous force between the plates is $\eta Au/d$ and this we may equate to the force ZqE acting on the macroion. If we then equate the zeta potential to the potential, $Zqd/\epsilon_r\epsilon_0 A$, across the condenser, we obtain the version of Smoluchovsky's equation relevant to these circumstances:

$$\nu = \zeta\epsilon_r\epsilon_0/\eta \tag{9.55}$$

This differs from equation (9.54) by the absence of the factor $\frac{2}{3}$. For intermediate values of the product κR, Henry showed that the factor $\frac{2}{3}$ in equation (9.54) should be replaced by $2X_1(\kappa R)/3$, where the function $X_1(\kappa R)$ increases from unity at low κR to a value of $\frac{3}{2}$ at high κR.

A more exact treatment expresses the zeta potential in terms of the charge as we have done in equation (9.53) and considers the structure of the ion atmosphere in more detail in terms of the Debye–Hückel theory. Exact expressions for the electrophoretic mobility of charged spheres have been given by Booth. These properly take into account the electrophoretic and relaxation effects, but are of considerable complexity.

The extension of these ideas to non-spherical particles presents grave problems which have not been satisfactorily solved. It is of interest, however, that Manning has presented a simplified account of the electrophoretic mobility of rod-like macroions, which is based on Kirkwood's theory of hydrodynamic interactions between arrays of spherical particles and on his own theory of counterion condensation, and he has quoted the result

$$\nu = \frac{4\pi\epsilon_r\epsilon_0 kT}{3\eta q} \{\ln{(\kappa b)}\}$$ (9.56)

This relation suggests that the mobility of the rod-like macroion is independent of the molecular weight and is proportional to the logarithm of the salt concentration. Both these predictions are approximately true in the case of DNA, though the calculated values of ν are about twice the experimental values. It must be understood, however, that these matters have not been settled to the satisfaction of all workers.

Hermans and Fujita have investigated the electrophoretic mobility of random-coil polyelectrolytes. They also predict that the mobility of high-molecular-weight coils in moderate ionic strengths should be independent of the molecular weight and depend only on the charge per monomer unit. This prediction has been verified for several systems.

Finally it is important to note that an uncharged macromolecule has an electrophoretic mobility of zero. Thus the mobility of proteins is zero at their isoelectric point.

9.17   Electrophoresis in gels

We have seen in chapter 6 that a cross-linked gel resembles a random three-dimensional net. Such a gel may be characterised by an average pore size, this being the average size of the holes in the net. It is possible to cause charged macromolecules to move in such gels under the influence of an electric field. If the size of the macroion is small compared with the pore size they will migrate with a mobility close to that in free solution, but if their size is comparable with or greater than the pore size they will be impeded in their motion. In general, the larger the macroion, the smaller will be its mobility in the gel.

For such reasons macroions migrate in gels with a velocity that depends on their molecular weight and size as well as their charge, even if they are rods or random coils, and the technique of gel electrophoresis has been of immense practical use in separating proteins and nucleic acids on the basis of both these parameters. This molecular-sieving effect has been observed in several types of gels including solubilised starch, agarose (another polysaccharide gel) and in cross-linked polyacrylamide. The latter has the advantage that the average pore size can be controlled over a wide range by altering the concentration of monomer that is employed in its preparation.

Unfortunately no adequate theory is available to describe electrophoresis in gels, but it is found empirically that the ratio of the mobility, ν, in the gel to that, ν_0, in free solution is roughly equal to the fraction, f_r, of the total gel volume that is available to the macroion in question. An expression for this fraction was discussed in section 6.6 so that from equation (6.26) we may write

$$\nu/\nu_0 = e^{-\alpha T s^2} \tag{9.57}$$

where T is the total concentration of polymeric material constituting the gel, α is a constant which depends on the degree of cross linking and s is a length characterising the radius of the domain occupied by the macroion. In accord with this, plots of ln (ν/ν_0) are frequently linear in T.

9.18   The viscosity of solutions of polyelectrolytes

When a solution of a polyelectrolyte is under shear, the stresses in the solvent in the vicinity of the macroions distort the ion atmosphere; since the counterions in the atmosphere are constrained to be close to the macroion, the solvent must also flow past them. This gives rise to an additional dissipation of energy so that the viscosity of the solution is higher than it would be if the macromolecule were to be uncharged. This is known as the *electroviscous effect*. At high ionic strengths, when the thickness of the ion atmosphere is small, the effect will be small, and the viscosity of the solution will be close to that which would be observed if the particle were uncharged.

Booth has derived a complex expression to describe the effect in the case of spherical particles, and this has been shown to yield results of the correct order of magnitude for several proteins. The effect is small, not exceeding 10%, and is reduced to negligible proportions by the addition of neutral salt to moderate concentrations (0.1 M NaCl). It is therefore desirable that moderate concentrations of salt be present in experiments designed to derive information concerning the shape of macroions from measurements of their intrinsic viscosity.

At high concentrations of polyelectrolyte a *second electroviscous effect* may be manifest, which originates from interactions between the charged particles. This effect is generally considered to be unimportant in practical circumstances.

With flexible macroions further complications arise, in that the conformation and size of the domain or of the equivalent sphere depend on the salt concentration, as we have seen in section 9.12. We have seen in section 8.3 that the reduced specific viscosity decreases with concentration of an uncharged macromolecule and extrapolates at zero concentration to the intrinsic viscosity. If the reduced specific viscosity is determined for a flexible polyelectrolyte in the absence of added salt it is commonly observed that it increases as the concentration decreases in the manner illustrated in figure 9.4. Only at very low concentrations does it fall in the conventional manner. This effect is to be attributed to the fact that, as the concentration of macroions is reduced, the concentration of counterions, and hence the ionic

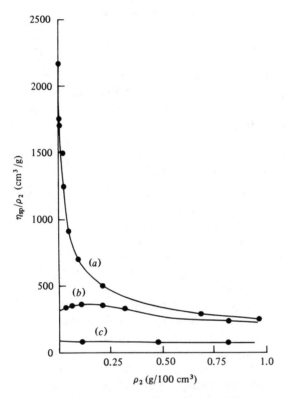

Figure 9.4. Plot of reduced specific viscosity, (η_{sp}/ρ_2), versus concentration, ρ_2, for polyvinylpiridinium bromide in various solvents: (a) pure water; (b) 0.001 M KBr in water; (c) 0.0335 M KBr in water. Note that the dramatic rise in η_{sp}/ρ_2 as the concentration is reduced, the ionic strength is reduced and the polymer expands. This effect is eliminated by the addition of neutral salt to the solution. This ensures that the ionic strength remains almost constant as the polymer is diluted so that it does not expand. (Data taken from R. M. Fuoss, *Disc. Faraday Soc.*, **11**, 127, (1951).)

strength, is reduced also. This means that the polymer chain expands as it is diluted and the notion of a well-defined conformation-dependent intrinsic viscosity becomes less meaningful. At infinite dilute, the macroion would be expected to be highly extended, and the intrinsic viscosity would correspond to this state of the polymer chain; however, such low concentrations are difficult to achieve in practice.

In order to mitigate this effect and to obtain intrinsic viscosities that appertain to the macromolecule in a defined conformational state it is desirable to measure the viscosities of a series of solutions of varying macroion concentration in which the ionic strength is kept constant by the addition of appropriate amounts of neutral salt. Alternatively the solvent may contain excess salt such that the contribution to the ionic strength by the counterions provided by the polyelectrolyte is negligible. With these precautions, plots of reduced specific viscosity versus macroion concentration resemble those obtained for uncharged polymers.

In moderate salt concentrations (0.1 M), highly charged flexible macroions are in a highly expanded state so that the expansion factor is expected to be proportional to $M_2^{0.1}$, which implies that the intrinsic viscosity is proportional to $M_2^{0.8}$ (see table 8.2). Only in the presence of high concentrations of salt such that theta conditions are approached does the exponent fall to a value close to 0.5. At very low salt concentrations the exponent may rise to about 1.8, indicating that the chain is fully stretched and rod like.

Equation (9.45) suggests that at low salt concentration the expansion factor is proportional to $c_3^{-0.2}$. This implies that the intrinsic viscosity should be proportional to $c_3^{-0.6}$. In practice lower values of the exponent are often observed.

9.19 The diffusion of polyelectrolytes

Let us first consider a solution containing a concentration gradient of a polyelectrolyte with no added salt. The macroions and counterions will tend to diffuse down the gradient as explained in section 8.13. If they were not charged, the small counterions would diffuse much faster than the large macroions and forge ahead down the gradient leaving the macroions behind. Since, however, they are charged, this would lead to a separation of electrical charges and a setting up of a gradient of electrical potential or an electrical field. This would tend to oppose the charge separation. Thus the counterions are held back and the macroions are accelerated. We may suppose that the field that is set up is just sufficient to cause both sorts of ion to diffuse at the same rate so that electrical neutrality is preserved at all points in the gradient. What would then be observed would be the diffusion of the neutral thermodynamic component, the polyelectrolyte.

In section 8.11 we argued that the driving force which caused the diffusional motion of molecules in a concentration gradient was the gradient of chemical potential. When we consider the diffusion of ions, we must add to this the electrostatic force due to the electric field set up in the manner described above.

Let us consider a solution containing a polyelectrolyte PX_Z and a neutral salt XY providing ions P, X and Y of valencies Z, -1 and $+1$ respectively. It will be convenient to specify the mole concentration, c_i, of each species, i, its frictional coefficient, f_i, its chemical potential, μ_i, and its charge number, z_i.

The total 'force' acting on each mole of component i is now seen to be $d\mu_i/dx + z_i qEN_A$, where E is the elctrical field strength. This force we will equate to $u_i f_i$, where u_i is the velocity. We may thus obtain three equations for the three ionic species, each of the form

$$u_i = (\mathrm{d}\mu_i/\mathrm{d}x + z_i q E N_A / N_A f_i) \tag{9.58}$$

We now make the assumption that there is no net flow of charge (no electric current); this will certainly be true as long as no electrodes connected to an external circuit are present in the solution. This assumption implies that

$$q c_X u_X - q c_Y u_Y - Z q c_P u_P = 0 \tag{9.59}$$

We may multiply the three expressions for the ionic velocities by the coefficients appearing in equation (9.59), sum the resulting expressions, equate the sum to zero in virtue of equation (9.59) and thus obtain an expression for qE. This in turn may be substituted into the expression for u_P to eliminate qE.

The result of these operations may be simplified by assuming that each point in the solution is electrically neutral so that

$$c_X = c_Y + Z c_P \tag{9.60}$$

and equating c_Y to c_3, the concentration of neutral salt; c_P to c_2, the concentration of neutral polyelectrolyte; $\mu_X + \mu_Y$ to μ_3, the chemical potential of the neutral salt; and $\mu_P + Z\mu_X$ to μ_2, the chemical potential of the neutral polyelectrolyte. The result is still moderately complicated.

To proceed further we may make the crucial assumption that there is no gradient in the chemical potential of the neutral salt. This will be true initially in a free boundary diffusion experiment provided the solvent used (see section 8.14) is in dialysis equilibrium with the solution of the polyelectrolyte. If we put $\mathrm{d}\mu_3/\mathrm{d}x = 0$ and equate $u_P(\mathrm{d}x/\mathrm{d}\ln c_2)$ to $-D_2$, as required by Fick's first law, we may finally obtain

$$D_2 = \frac{RT}{N_A f_P} \left(1 + \rho_2 \frac{\mathrm{d}\ln y_2}{\mathrm{d}\rho_2}\right) \bigg/ \left\{1 + \frac{z^2 c_2}{f_P[c_3(f_Y^{-1}) + (f_Y^{-1}) + Z c_2 f_X^{-1}]}\right\} \tag{9.61}$$

It is readily seen from this that, provided there is some added salt ($c_3 \neq 0$), the value of D_2 extrapolated to infinite dilution ($c_2 = 0$) is equal to RT/Nf_P. In so far as f_P can be related to the size and shape of the macroion, measurements of D_2^0 will provide information about these characteristics.

We may also see from equation (9.61) that in the complete absence of added salt such that $c_3 = 0$, D_2^0 is equal to $RT/N_A(f_P + Zf_X)$. That is to say that the effective frictional coefficient is now that of the macroion plus those of all its counterions.

It is important to note that the treatment that we have just outlined is by no means rigorous. In particular the frictional coefficients that we have used are not well defined. We might suppose for instance that there would be some measure of hydrodynamic interaction between the macroion and its counterions similar in nature to the electrophoretic effect mentioned in section 9.15, and that the magnitude of the effect would depend on the thickness of the ion atmosphere.

More exact treatments using the methods of irreversible thermodynamics and based on equation (8.25) have been produced. In such treatments equations similar in form to equation (8.35) result. In these equations the diffusion coefficients and cross-diffusion coefficients depend on the parameters which describe the hydrodynamic interactions between the various species present. A detailed discussion of these matters is beyond the scope of this book.

9.20   The sedimentation of polyelectrolytes

We saw in the previous section that the low diffusion coefficient of the macroion relative to the counterions gave rise to a potential gradient that tended to accelerate the former. In the ultracentrifuge, in circumstances in which sedimentation is the only transport process, it is the heavy macroion which tends to forge ahead leaving its counterions behind. This sets up a potential gradient known as the *sedimentation potential* which acts so as to retard the macroion and accelerate the counterions. This effect is known as the *primary charge effect* in the context of the sedimentation of polyelectrolytes.

Another effect, the *secondary charge effect*, arises if the (small) sedimentation coefficients of the counterion and coion differ. In this case the heavier of the two forges ahead of the lighter thus setting up an additional potential gradient. This additional potential gradient in turn accelerates or retards the macroion, according to whether the counterion or coion sediments the faster.

Both the primary and secondary charge effects may be treated in a manner analogous to that deployed in the previous section. In the absence of concentration gradients the total force acting on each ion is $M_i(1 - \bar{v}_i\rho) \omega^2 x + z_i qE$, so that the analogue of equation (9.58) is

$$u_i = \{\omega^2 x M_i(1 - \bar{v}_i\rho)/N_A + z_i qE\}/f_i \tag{9.62}$$

Equations (9.59) and (9.60) are still valid in the present context so that the electric field E may be eliminated from the expression for u_p in a similar manner.

We may put $M_X + M_Y$ equal to M_3, the molar mass of the salt, and $M_X\bar{v}_X + M_Y\bar{v}_Y$ equal to $M_3\bar{v}_3$, the partial molar volume of the salt, so that $M_X(1 - \bar{v}_X\rho) + M_Y(1 - \bar{v}_Y\rho) = M_3(1 - \bar{v}_3\rho)$. Similarly $M_P(1 - \bar{v}_P\rho) + ZM_X(1 - \bar{v}_X\rho) = M_X(1 - \bar{v}_2\rho)$. Finally we equate $u_p/\omega^2 x$ to s_p the sedimentation coefficient of the macroion and express $1/s_P$ as a power series in ρ_2. Neglecting terms in ρ_2^2 and above we find:

$$\frac{1}{s_P} = \frac{N_A f_P}{M_2(1 - v_2) - \dfrac{ZM_3(1 - \bar{v}_3\rho)}{1 + f_Y/f_X}} + Q\rho_2 \qquad (9.63)$$

where Q is a complicated function of several of the parameters excluding ρ_2.

The term f_Y/f_X accounts for the secondary charge effect; if $f_Y = f_X$, the effect is eliminated. This may be done in practice by choosing a counterion and a coion with closely similar hydrated radii; Na^+ and Cl^- are good contenders and are frequently employed for this purpose.

s_P may in principle be measured in the manner discussed in section 8.19 at a series of concentration of the polyelectrolyte. The value of $1/s_P$ extrapolated to $\rho_2 = 0$ is then given when $f_Y = f_X$ by

$$1/s_P^0 = \frac{N_A f_P}{M_2(1 - v_2\rho) - \frac{1}{2}ZM_3(1 - v_3\rho)} \qquad (9.64)$$

We see that the effect of the charge on the macroion is still present at infinite dilution of the macroion and is not eliminated even at infinite salt concentration. However Eisenberg has argued that the denominator of the right-hand side of equation (9.64) may be equated to $M_2(\partial\rho/\partial\rho_2)_{\mu_3}$ to give

$$s_P^0 = M_2(\partial\rho/\partial\rho_2)_{\mu_3}/N_A f_P \qquad (9.65)$$

The quantity $(\partial\rho/\partial\rho_2)_{\mu_3}$ is the same quantity that was discussed in section 9.4 and is obtainable by equilibrium dialysis experiments. It should be noted that M_2 is the molar mass of the neutral polyelectrolyte, whereas f_P is the frictional coefficient of the macroion. It should also be noted that this treatment does not take any account of the variation of f_P with salt concentration, either as a result of conformation changes of a flexible macroion or because of hydrodynamic interactions between macroion and counterion. The latter effects are formally allowed for in more extended treatments based on irreversible thermodynamics and which result in cross sedimentation coefficients describing such interactions.

In the complete absence of added salt ($c_3 = 0$) it may be shown that the treatment outlined above leads to $s_P = M_2(1 - \bar{v}_2\rho)/(f_P + Zf_X)N_A$. That is to say the relevant frictional coefficient is that of the macroion plus that of all of its counterions. Under these circumstances s_P will be low.

If the frictional coefficients of the macroion in diffusion and sedimentation are assumed to be identical (and this may not be so), f_P can be eliminated from the limiting expressions for D_2^0 and s_P^0 to give an expression analogous to the Svedberg equation (equation (8.46)) which takes the form

$$M_2 = \frac{RTs_P^0}{D_2^0(\partial\rho/\partial\rho_2)_{\mu_3}}$$ (9.66)

Here it should be noted that M_2 is the molar mass of the neutral polyelectrolyte together with its counterions, if ρ_2 refers similarly to the mass concentration of the neutral polyelectrolyte. Alternatively M_2 may be taken as the molar mass of the macroion provided ρ_2 is the mass concentration of this species.

It should finally be noted that equation (9.66) is also valid if there is no added salt. In this case $(\partial\rho/\partial\rho_2)$ is exactly equal to $(1 - \bar{v}_2\rho)$.

Problems

9.1 A solution of RNA in which the nucleotide residue concentration is 10^{-5} mol/cm^3 is dialysed exhaustively against a solution of NaCl at a concentration of 10^{-6} mol/cm^3 and containing a very low concentration of buffer sufficient to maintain the pH at 6.0. What is the concentration of Na$^+$ ions in the RNA solution at equilibrium? What is the pH of the RNA solution? How would the pH of the RNA solution change if the NaCl concentration in the outer solution were increased to 10^{-4} mol/cm^3?

9.2 Estimate the osmotic pressure at 25°C of solutions of 5S RNA at concentrations of 0.5, 1.0 and 5.0 mg/cm^3 in 0.1 M NaCl. Assume that the charge number of the RNA is -120 and its molecular weight 40 000. Assume that 5S RNA is a compact sphere of negligible volume. What difficulties do you foresee in using osmotic pressure to measure the molecular weight of 5S RNA accurately under these conditions? How could these difficulties be reduced?

9.3 Calculate the potential difference across the membrane separating a solution of a negatively charged macroion in Donnan equilibrium with a buffer solution when the pH difference between the solutions is 1.0 and the temperature is 20°C.

9.4 Calculate the ionic strength of a solution containing 0.01 M MgCl$_2$, 0.1 M Na$_2$SO$_4$, 0.001 M acetic acid and 0.001 M sodium acetate.

9.5 Calculate the energy required to bring one hydrogen ion up to the surface of each of the molecules in one mole of a globular protein. The protein molecule has a radius of 1 nm and carries a charge equal to that of 15 protons. Assume that the charge is smeared over the surface of the spherical macromolecule and that the ionic strength is 0.1 M. The radius of the H$^+$ ion should be ignored.

9.6 The rod-like triple-helical complex of polyadenylic acid with two molecules of polyuridylic acid carries three electronic charges per 0.34 nm along its length. Calculate the number of Na^+ ions that would condense onto a molecule 50 nm long. Assume that the temperature is 20°C, the solvent is water with a dielectric constant of 80 and that the only small ions present are Na^+ and Cl^-.

9.7 Estimate the expansion factor, α, for a fully ionised molecule of polyphosphoric acid of degree of polymerisation 400 dissolved in 0.035 M NaBr. Assume that the radius of gyration of the unperturbed polymer is 3.5 nm. Base your calculations on the Flory theory and neglect expansion effects that would still be present if the molecule were uncharged. Experiment suggests a value of 1.8; how would you account for any discrepancy between your value and this.

Assume throughout these problems that:
  The charge on a proton is 1.602×10^{-19} C
  Avogadro's number is 6.02×10^{23}/mol
  The gas constant is 8.314 J/K mol
  The Faraday constant is 9.65×10^4 C/mol
  The permittivity of free space is 8.85×10^{-12} F/m
  The dielectric constant of water at 20°C is 80

Appendixes

Appendix A. Derivation of the partition function for an ensemble of partially helical polymer chains using the matrix method

Equations (5.10) may be simply written in matrix notation as

$$\begin{pmatrix} Q_{n+1,\,c} \\ Q_{n+1,\,h} \end{pmatrix} = T \begin{pmatrix} Q_{n,\,c} \\ Q_{n,\,h} \end{pmatrix} \tag{A.1}$$

where T is the 2×2 transformation matrix defined by

$$T = \begin{pmatrix} 1 & 1 \\ s\sigma & s \end{pmatrix} \tag{A.2}$$

We may readily see by induction that

$$\begin{pmatrix} Q_{n,\,c} \\ Q_{n,\,h} \end{pmatrix} = T^{n-1} \begin{pmatrix} Q_{1,\,c} \\ Q_{1,\,h} \end{pmatrix} \tag{A.3}$$

We may also note that $Q_n = Q_{n,c} + Q_{n,h}$ so that

$$Q_n = (1 \quad 1) \begin{pmatrix} Q_{n,\,c} \\ Q_{n,\,h} \end{pmatrix} \tag{A.4}$$

and that if $Q_{1,c} = 1$ and $Q_{1,h} = s$

$$\begin{pmatrix} Q_{1,\,c} \\ Q_{1,\,h} \end{pmatrix} = T \begin{pmatrix} 0 \\ 1 \end{pmatrix} \tag{A.5}$$

It then follows that

$$Q_n = (1 \quad 1)\, T^n \begin{pmatrix} 0 \\ 1 \end{pmatrix} \tag{A.6}$$

To evaluate the matrix product, T^n, we assume that T may be diagonalised

243

and written in the form

$$T = \Lambda \begin{pmatrix} \lambda_1 & 0 \\ 0 & \lambda_2 \end{pmatrix} \Lambda^{-1} \tag{A.7}$$

where the product of the two 2×2 matrices, $\Lambda \Lambda^{-1}$ is the 2×2 unit matrix and λ_1 and λ_2 are the eigenvalues of T. We then see that

$$Q_n = (1 \quad 1) \Lambda \begin{pmatrix} \lambda_1^n & 0 \\ 0 & \lambda_2^n \end{pmatrix} \Lambda^{-1} \begin{pmatrix} 0 \\ 1 \end{pmatrix} \tag{A.8}$$

The components of the matrices Λ and Λ^{-1} may be expressed in terms of λ_1 and λ_2 as

$$\Lambda = \begin{pmatrix} 1 & -1 \\ -(1 - \lambda_1) & (1 - \lambda_2) \end{pmatrix}$$

and

$$\Lambda^{-1} = \begin{pmatrix} (1 - \lambda_2)/(\lambda_1 - \lambda_2), & 1/(\lambda_1 - \lambda_2) \\ (1 - \lambda_1)/(\lambda_1 - \lambda_2), & 1/(\lambda_1 - \lambda_2) \end{pmatrix} \tag{A.9}$$

Substituting equations (A.9) into (A.8) and multiplying out the matrix products then gives equation (5.10). The reader may readily verify the truth of equation (A.7) and that the product $\Lambda \Lambda^{-1}$ gives the unit matrix by direct multiplication.

The eigenvalues, λ_1 and λ_2, are readily found by solving the quadratic equation in λ given by

$$\begin{vmatrix} 1 - \lambda & 1 \\ s\sigma & s - \lambda \end{vmatrix} = 0 \tag{A.10}$$

The solution is represented by equation (5.11).

Appendix B.   The representation of waves by complex numbers

The complex number $a\,e^{i\theta}$ is identically equal to $a \cos \theta + ia \sin \theta$, so that the real part, $a \cos \theta$, is equal to the real part of $a\,e^{i\theta}$. Thus a plane monochromatic wave represented by $E \cos (\omega t - kx)$ is also represented by the real part of $E\,e^{i(\omega t - kx)}$.

It then follows that the resultant of a number of waves polarised in the same plane may be represented either by $\Sigma_i\,E_i \cos (\omega t - kx + \phi_i)$ or by the real part of $\Sigma_i\,E_i\,e^{i(\omega t - kx + \phi_i)}$, where E_i and ϕ_i are the amplitudes and phase angles of the individual components.

If we further suppose that the resultant wave has an amplitude, E_r, and phase, ϕ_r, we may set

$$E_r \, e^{i(\omega t - kx + \phi_r)} = \sum_i E_i \, e^{i(\omega t - kx + \phi_i)} \tag{B.1}$$

The equality of the real part of the left-hand and right-hand sides of this equation ensures that $E_r \cos(\omega t - kx + \phi_r) = \Sigma_i E_i \cos(\omega t - kx + \phi_i)$. If we divide both sides of equation (B.1) by $e^{i(\omega t - kx)}$ we obtain

$$E_r \, e^{i\phi_r} = \sum_i E_i \, e^{i\phi_i} \tag{B.2}$$

The complex conjugate of a complex number is defined to be the complex number obtained by replacing i by $-$i. If two complex numbers are equal, their complex conjugates are also equal, so that if $a + ib = c + id$ then also $a - ib = c - id$. Thus we may multiply both sides of equation (B.2) by their respective complex conjugates to obtain

$$E_r^2 = \left(\sum_i E_i \, e^{i\phi_i} \right) \left(\sum_i E_i \, e^{-i\phi_i} \right) \tag{B.3}$$

E_r^2 is proportional to the intensity of the resultant wave and the right-hand side may be written as $\Sigma_i \Sigma_j E_i E_j \, e^{i(\phi_i - \phi_j)}$. For every term in $\phi_i - \phi_j$ in the double summation there is a term in $\phi_j - \phi_i$, so that the expression may be written in the form $\frac{1}{2} \Sigma_i \Sigma_j E_i E_j [e^{i(\phi_i - \phi_j)} + e^{-i(\phi_i - \phi_j)}]$.

Now it may be shown that $\frac{1}{2}(e^{i\theta} + e^{-i\theta})$ is equal to $\cos \theta$, so that from equation (B.3) we find

$$E_r^2 = \sum_i \sum_j E_i E_j \cos(\phi_i - \phi_j) \tag{B.4}$$

which is a real number as required of an experimentally measurable quantity.

The phase, ϕ_r, of the resultant may be found by equating the real and imaginary parts of both sides of equation (B.2) to obtain

$$E_r \cos \phi_r = \sum_i E_i \cos \phi_i$$

$$E_r \sin \phi_r = \sum_i E_i \sin \phi_i \tag{B.5}$$

On taking ratios we thus find

$$\tan \phi_r = \sum_i E_i \sin \phi_i \Big/ \sum_i E_i \cos \phi_i \tag{B.6}$$

so that the phase too is real.

It is sometimes convenient to refer to the complex number, $E_r \, e^{i\phi_r}$, as the 'amplitude' of the wave $E_r \cos(\omega t - kx + \phi_r)$, meaning that the intensity of the wave (proportional to E_r^2) is proportional to the 'amplitude' times its complex conjugate. This usage should not result in confusion if an amplitude noted as being complex is understood in this way.

Appendix C.   The field autocorrelation function

The form of Fick's second law that describes the diffusion of a solute molecule in three dimensions is

$$\frac{\partial \rho}{\partial t} = D\left[\frac{\partial^2 \rho}{\partial x^2} + \frac{\partial^2 \rho}{\partial y^2} + \frac{\partial^2 \rho}{\partial z^2}\right] \tag{C.1}$$

Solutions to this partial differential equation specify the concentration at a point x, y, z at time t. The concentration within a small volume element is proportional to the probability of finding any particular molecule therein. Thus the probability of finding a molecule in a small volume element at x, y, z at time t after it was known to be at the origin is governed by the same differential equation. If P is this probability we have

$$\frac{\partial P}{\partial t} = D\left[\frac{\partial^2 P}{\partial x^2} + \frac{\partial^2 P}{\partial y^2} + \frac{\partial^2 P}{\partial z^2}\right] \tag{C.2}$$

This equation may be transformed to spherical polar coordinates. If this is done and it is noted that P is independent of direction and depends only on the radial distance, s, from the origin, we obtain

$$\frac{\partial P}{\partial t} = D\frac{1}{s^2}\frac{\partial}{\partial s}\left[s^2\frac{\partial P}{\partial s}\right] \tag{C.3}$$

where P is a function only of s and t and represents the probability of finding a molecule in any small volume element at a radial distance s from the origin at time t, given that it was at the origin at $t = 0$. The probability of finding the molecule within the spherical shell of radius, r, and thickness, dr, is then equal to $4\pi s^2 P\, ds$. Clearly the integral of this from $s = 0$ to $s = \infty$ must be equal to unity since the molecule must be, with probability unity, at some distance from the origin at all times.

The solution to equation (C.3), which meets this boundary condition, is

$$P = \frac{1}{(4\pi Dt)^{3/2}}\, e^{-s^2/4Dt} \tag{C.4}$$

as may be readily verified.

The expression for the field autocorrelation function, equation (7.26), may be written

$$G^1(t) = \int_0^\pi \int_0^\infty P\, e^{i(\omega t + \boldsymbol{h}\cdot\boldsymbol{s})}\, 4\pi s^2\, \sin\theta\, d\theta\, ds \tag{C.5}$$

In this expression θ is the angle between a radius vector s and the scattering vector h, so that $4\pi s^2 \sin \theta \, d\theta \, ds$ is the total volume of space which lies between radii s and $s + ds$ and between angles θ *and* $\theta + d\theta$ to the direction of h. The vector product $h \cdot s$ is equal to $hs \cos \theta$, so that the integration over θ is easily performed to yield

$$G^1(t) = \frac{8 \, e^{i\omega t} \pi}{(4\pi Dt)^{3/2} h} \int_0^\infty e^{-s^2/4Dt} s \sin (hs) \, ds \qquad (C.6)$$

where we have substituted from equation (C.4) for P.

The integral over s in equation (C.6) may be readily evaluated by parts to give

$$G^1(t) = 2 \, e^{i\omega t} \, e^{-h^2 Dt} \qquad (C.7)$$

as indicated by equation (7.27).

Appendix D. One-dimensional flow in isothermal systems

In this appendix a schematic derivation of equation (8.25), is presented.

The starting point is Gibbs' fundamental equation which relates changes in the internal energy, dU, entropy, dS, and number of moles of component i, dn_i, for a system maintained at equilibrium:

$$dU = T \, dS + \sum_i (\mu_i + M_i \phi_i) \, dn_i \qquad (D.1)$$

where ϕ_i is the potential energy per unit mass of component i acted on by external forces, F_i. We may recollect that $F_i = -\partial \phi_i/\partial x$.

We now apply this relation to a small volume element of the solution at position x and consider the energy and entropy changes per unit volume within it, du and ds. Although irreversible flows are assumed to be taking place within it, we may suppose that its interactions with its surroundings occur reversibly if it is small enough so that equation (D.1) applies. Dividing all through by an element of time, dt, and by the volume of the element, we obtain

$$\frac{du}{dt} = T \frac{ds}{dt} + \sum_i (\mu_i/M_i + \phi_i) \frac{d\rho_i}{dt} \qquad (D.2)$$

We next suppose that there are flows of heat, J_q, and of component i, J_i, into and out of the element. We also consider the total flow of energy, J_e. To proceed we apply the continuity equations based on the conservation of mass and of energy to J_i and J_e respectively.

The flow of energy, J_e, may be written as $J_q + \Sigma_i (\mu_i/M_i + \phi_i) J_i$, since energy is transported either as heat or as a flow of matter. If we apply the equation of continuity to this we obtain

$$\left(\frac{\partial u}{\partial t}\right)_x = -\left(\frac{\partial J_q}{\partial x}\right)_t - \sum_i \frac{\partial}{\partial x}\left[J_i(\mu_i/M_i + \phi_i)\right]_t \tag{D.3}$$

Similarly if we apply the equation of continuity to the flow of component i we obtain

$$\left(\frac{\partial \rho_i}{\partial t}\right)_x = -\left(\frac{\partial J_i}{\partial x}\right)_t \tag{D.4}$$

The application to the flow of heat requires more care. Heat is not a function of state; that is to say it is not something about which we can say that a certain amount is contained in a system. Instead we must consider the entropy. If the flow of heat at the boundary of a system at which the temperature is T is J_q, we can talk about the flow of entropy into the system across this boundary owing to its interaction with its surroundings. This flow of entropy is equal to J_q/T. If we apply the equation of continuity to this entropy flow, the increase of entropy within the system corresponds to the term Δs_e defined in equation (8.20). We thus see that

$$\left(\frac{\partial s_e}{\partial t}\right)_x = -\frac{\partial}{\partial x}\left(\frac{J_q}{T}\right)_t \tag{D.5}$$

Finally we note that the rate of generation of entropy within the system due to irreversible processes occurring therein, θ, is given directly by equation (8.21) as:

$$\theta = \frac{ds}{dt} - \frac{ds_e}{dt} \tag{D.6}$$

From equations (D.2) to (D.6) it is then easy to show that

$$\theta = J_q \frac{\partial}{\partial x}\left(\frac{1}{T}\right) - \frac{1}{T}\sum_i J_i \frac{\partial}{\partial x}(\mu_i/M_i + \phi_i)_t \tag{D.7}$$

For isothermal systems, $\partial(1/T)/\partial x = 0$ and the first term of (D.7) drops out. The second term is of the form $\sum_i J_i X_i/T$, where we may identify the thermodynamic force, X_i, with $\partial(\mu_i/M_i + \phi_i)/\partial x$. This we may write in the form

$$X_i = F_i - \frac{1}{M_i}\left(\frac{\partial \mu_i}{\partial x}\right)_{t,T} \tag{D.8}$$

This completes the application of irreversible thermodynamics to the problem. To proceed we first note that the chemical potential of a component is a function of the pressure, P, and of the concentration of all of the components. However the chemical potentials of the entire set of components are not independent of each other. This means that we need not include the dependence of μ_i on the concentration of one of them. We chose to omit the dependence on the concentration of the solvent. With these considerations in mind we may then write formally

$$\left(\frac{\partial \mu_i}{\partial x}\right)_T = \left(\frac{\partial \mu_i}{\partial P}\right)_T \left(\frac{\partial P}{\partial x}\right)_t + \sum_{i=2}^{n} \left(\frac{\partial \mu_i}{\partial \rho_j}\right)_{P,T} \left(\frac{\partial \rho_j}{\partial x}\right)_t \qquad (D.9)$$

The term $\partial \mu_i / \partial P$ is readily shown to be equal to $\overline{V}_i$ and $\partial \rho_j / \partial x$ is the concentration gradient g_j of component j. For simplicity we will write $\partial \mu_i / \partial \rho_j$ as μ_{ij}. It remains to investigate $\partial P / \partial x$.

If we consider the mechanical forces acting on a volume element of solution (see figure 8.12) we see that a gradient of the pressure in its vicinity produces an upwards force dP/dx. On the other hand, the external forces acting on the various components within produce a downwards force equal to $\sum_i \rho_i F_i$. We now assume that the centre of mass of the volume element is at mechanical equilibrium and undergoes no acceleration. It then follows that

$$\frac{dP}{dx} = \sum_i \rho_i F_i \qquad (D.10)$$

Having made this assumption we have tacitly committed ourselves to measuring flows relative to the centre of mass of the volume element. If we substitute from equations (D.9) and (D.10) into equation (D.8) we would obtain an expression for the thermodynamic force relevant to flows, J_i^m, defined in this way. We may denote such flows by J_i^m and the corresponding forces by X_i^m. However, we require an expression for the forces for flows relative to the centre of volume, which we denote by J_i^*. We require to transform our equation to obtain expressions for X_i^* given the expression for X_i^m such that $\sum_i J_i^m X_i^m = \sum_i J_i^* X_i^*$. This can be done in a straightforward manner, albeit after considerable algebraic manipulation, if we note that the velocity of the centre of volume is given as $u^* = \sum_i \overline{v}_i \rho_i u_i$ and that of the centre of mass by $u^m = \sum_i \rho_i u_i$, where u_i is the velocity of a component i relative to the cell.

Since the flows relative to the centre of volume or centre of mass are not independent of each other, we eliminate from consideration the flow of the solvent. This is facilitated by invoking the Gibbs–Duhem equation (3.15). When this programme of mathematical manipulations is carried through, there results equation (8.25).

Appendix E. Some useful mathematical formulae

Integrals

$$\int_0^\infty e^{-ax^2} \, dx = \frac{1}{2} \sqrt{\frac{\pi}{a}}$$

$$\int_0^\infty e^{-ax^2} x \, dx = \frac{1}{2a}$$

$$\int_0^\infty e^{-ax^2} x^2 \, dx = \frac{1}{4} \sqrt{\frac{\pi}{a^3}}$$

$$\int_0^\infty e^{-ax^2} x^3 \, dx = \frac{1}{2a^2}$$

Series expansions

$$(1 + x)^{-1} = 1 - x + x^2 - x^3 \cdots$$

$$(1 - x)^{-1} = 1 + x + x^2 + x^3 \cdots$$

$$\ln(1 + x) = x - x^2/2 + x^3/3 \cdots$$

These are valid for $-1 < x < 1$.

$$e^x = 1 + x + x^2/2 \cdots$$

Summation of series

$$\sum_{i=1}^n id = d + 2d + 3d + 4d + \cdots + nd = dn(n + 1)/2$$

$$\sum_{i=1}^n r^i = r + r^2 + r^3 + \cdots + r^n = (r^{n+1} - r)/(r - 1)$$

An approximation to $\sum_{i=1}^n (1/i)$ for large n is given by $\ln n + 1/2n + 0.577$ (the constant 0.577 is Euler's constant).

An approximation to $\ln(n!) = \sum_{i=1}^n \ln(i)$ for large n is given by Stirling's approximation as $n \ln(n) - n$.

Answers to problems

1.1 171 429
233 333
300 000

1.3 0.86×10^6

2.3 14 kJ/mol

3.1 156 500
2.32×10^{-5} kg^{-2}m^3 mol
1.13 m^3/mol
0.12 m^3/mol

3.5 33.7°C

3.6 2.45×10^{-4} cm^3 mol/g^2

3.7 1.13×10^8

4.1 254 nm
6.77
0.59 nm

4.3 9.48 nm

4.4 298 nm
0.65 nm

4.5 636 nm
193 nm
4.8 nm

4.6 5.0×10^{-5}
1.6×10^{-6}

4.8 2.5

5.2 −0.0269 kJ/K mol
−0.0256 kJ/K mol

5.5 1.16 kJ/K mol
0.0091 kJ/K mol

6.1 1.15 N/cm
0.37 N/cm

6.2 133 nm
1.16

7.2 2.64 nm

8.1 4070 S

8.2 8521 rev/min

8.3 3.66×10^{-15} N
1.88×10^{-10} N/s/m

8.5 1.54

8.6 4.2 S

8.7 324 000

8.8 53 cm^3/g

8.9 54.0
57.0
67.3
69.2
92.3
152.0
163.0

9.1 1.01×10^{-5} mol/cm^3

9.2   44.9 Pa
       117.6 Pa
       170.2 Pa

9.3   0.058 V

9.4   0.331 M

9.5   12.8 kJ/mol

9.6   371

9.7   5.5

FURTHER READING

Chapter 1

Proteins

1.1 R. E. Dickerson & I. Geis, *The Structure and Action of Proteins*, (W. A. Benjamin, New York, 1969).
A useful and clear introduction to the structure and biochemistry of proteins.

1.2 H. Neurath & R. L. Hill, *The Proteins*, 3rd ed, 3 volumes, (Academic Press, New York, 1977).
A compendium of articles on various aspects of proteins. Previous editions of this series also contain useful articles.

Nucleic acids

1.3 R. L. P. Adams, R. H. Burdon, A. M. Campbell & R. M. S. Smellie, *Davidson's The Biochemistry of the Nucleic Acids*, 8th ed, (Chapman & Hall, London, 1976).
A revised edition of a classical introduction to the chemistry and biochemistry of nucleic acids.

1.4 V. A. Bloomfield, D. M. Crothers & I. Tinoco, *Physical Chemistry of Nucleic Acids*, (Harper & Row, New York, 1974).
An authoritative account of many aspects of nucleic acid physical chemistry.

Polysaccharides

1.5 E. A. Davidson, *Carbohydrate Chemistry*, (Holt, New York, 1967).
A general survey.

Colloids

1.6 K. J. Mysels, *Introduction to Colloid Chemistry*, (Interscience, New York, 1959).

Synthetic Polymers

1.7 F. W. Billmeyer, *Textbook of Polymer Science*, 2nd ed, (Wiley–Interscience, New York 1970).
A general account of synthetic polymers.

1.8 D. B. V. Parker, *Polymer Chemistry*, (Applied Science Publishers, London, 1974).
A clear basic text on polymers.

1.9 P. J. Flory, *Principles of Polymer Chemistry*, (Cornell University Press, Ithaca, 1953).
A classical text covering many spects of polymers.

1.10   F. A. Bovey, *Polymer Conformation and Configuration*, (Academic Press, New York, 1969)
An interesting review of stereochemical aspects of polymers including stereo-regular polymers.

Molecular-weight distribution and averages

1.11   C. Tanford, *Physical Chemistry of Macromolecules*, (Wiley, New York, 1961)
Chapter 3 contains a useful discussion of molecular-weight distributions and averages.

Chapter 2

Intermolecular forces

2.1   M. Davies, *Some Electrical and Optical Aspects of Molecular Behaviour*, (Pergamon, Oxford, 1965).
An elementary and readable account.

2.2   E. A. Moelwyn-Hughes, *Physical Chemistry*, (Pergamon, Oxford, 1957).
A detailed but not excessively mathematical treatment of several aspects.

2.3   *Discussions of the Faraday Society*, No. 40, (1965).
This volume is devoted to intermolecular forces. Note especially the articles by H. C. Longuet-Higgins (p. 7) and A. D. McLachlan (p. 239).

2.4   J. N. Isrealachvili, *Quart. Rev. Biophys.*, 6, 341, (1973).
A review from a more modern standpoint.

2.5   J. O. Hirschfelder, C. F. Curtiss & R. B. Bird, *Molecular Theory of Gases and Liquids*, (Wiley, New York, 1954).
A classical work but somewhat mathematical.

Water structure and hydrophobic bonds

2.6   D. Eisenberg & W. Kauzman, *The Structure and Properties of Water*, (Clarendon Press, Oxford, 1969).
A very clear account of many aspects of water.

2.7   J. L. Kavanu, *Water and Solute-Water interactions*, (Holden Day, San Francisco, 1964).
Terse but non-mathematical.

2.8   C. Tanford, *The Hydrophobic Effect*, (Interscience, New York, 1973).
A clear account of the solubility of hydrocarbons etc. in water and of the formation of micelles.

Chapter 3

Thermodynamics and statistical mechanics

3.1   W. J. Moore, *Physical Chemistry*, 3rd ed, (Prentice Hall, Englewood Cliffs, N.J., 1963).
This standard work on physical chemistry contains detailed explanations of classical thermodynamics.

3.2   E. A. Guggenheim, *Thermodynamics*, 5th ed, (North Holland, Amsterdam, 1967).
An authoritative and lucid book but not suitable as an introduction.

3.3   R. W. Gurney, *Introduction to Statistical Mechanics*, (McGraw–Hill, New York, 1949).
Suitable for a beginner.

3.4 T. L. Hill, *An Introduction to Statistical Mechanics*, (Addison–Wesley, Reading Mass., 1971).
Somewhat mathematical but with a clear discussion of the fundamentals and useful chapters on polymers and solutions.

3.5 A. E. Guggenheim, *Mixtures*, (Oxford University Press, Oxford, 1952).
A lucid but advanced discussion of lattice theories of mixtures.

Thermodynamics of polymers

3.6 P. J. Flory, *Principles of Polymer Science*, (Cornell University Press, Ithaca, 1953).
Authoritative treatment written clearly and without advanced mathematics.

3.7 H. Tompa, *Polymer Solutions*, (Butterworths, London, 1956).
A useful discussion.

3.8 P. J. Flory, *J. Chem. Phys.*, **10**, 51–61, (1942).
Flory's first presentation of the Flory–Huggins theory of concentrated polymer solutions.

3.9 M. L. Huggins, *Ann. N.Y. Acad. Sci.*, **43**, 1–32, (1942); *J. Am. Chem. Soc.*, **46**, 151–158, (1942); *J. Am. Chem. Soc.*, **64**, 1712–1719, (1942).
Huggins' first presentation of the Flory–Huggins theory.

3.10 P. J. Flory & W. R. Krigbaum, *J. Chem. Phys.*, **18**, 1086–1094, (1950).
First presentation of the Flory–Krigbaum theory of dilute solutions of polymers.

3.11 D. K. Carpenter, *Solution Properties*, in *Encyclopedia of Polymer Science and Technology* (ed: H. F. Mark, N. G. Gaylord & D. B. Bikales) vol. 12, (Interscience, New York, 1970).
An account of more recent developments.

3.12 K. E. Van Holde, *Physical Biochemistry*, (Prentice Hall, Englewood Cliffs, N.J., 1971).
An excellent introduction for the non-specialist.

Fluctuation spectroscopy

3.13 M. Weissman, H. Schindler & G. Feher, *Proc. Nat. Acad. Sci. USA*, **73**, 2776–2780, (1976).
A method for measuring the molecular weight of DNA.

Chapter 4

4.1 C. Tanford, *Physical Chemistry of Macromolecules*, (Wiley, New York, 1961).
Many of the theorems mentioned in this chapter are derived in a simple manner. Recommended.

4.2 P. J. Flory, *Statistical Mechanics of Chain Molecules*. (Interscience, New York, 1969).
An excellent and lucid description of most matters covered by this chapter. Requires an understanding of matrix algebra for full comprehension.

4.3 M. Volkenstein, *Configurational Statistics of Polymeric Chains*, (Translated from the Russian edn, S. N. Timasheff & M. J. Timasheff) (Interscience, New York, 1963).
A classic, but mathematical and not for the beginner.

4.4 T. M. Birshtein & O. B. Ptitsyn, *Conformations of Macromolecules*, (Translated from the Russian edn, S. N. Timasheff & M. J. Timasheff) (Interscience, New York, 1966).
Another classic.

4.5   V. A. Bloomfield, D. M. Crothers & I. Tinoco, *Physical Chemistry of Nucleic Acids*, (Harper & Row, New York, 1974).
General introduction with particular emphasis on nucleic acids.
4.6   A. J. Hopfinger, *Conformational Properties of Macromolecules*, (Academic Press, New York, 1973).
Tersely written and quite mathematical; some emphasis on polypeptides; many useful tabulations of data.

Chapter 5

Helix-coil transitions
5.1   P. J. Flory, *Statistical Mechanics of Chain Molecules*, (Wiley, New York, 1969).
Chapter 5 contains a clear exposition of helix-coil transition theory.
5.2   D. A. Poland & H. A. Scheraga, *Theory of Helix-Coil Transitions in Biopolymers*, (Academic Press, New York, 1970).
Mathematical and exhaustive.

Protein denaturation
5.3   S. Lapanje, *Physicochemical Aspects of Protein Denaturation*, (Interscience, New York, 1978).
Covers most aspects.

Chapter 6

Deformation of gels and chain expansion
6.1   P. J. Flory, *Principles of Polymer Chemistry*, (Cornell University Press, Ithaca, 1953).
A classic, lucidly written.
6.2   T. L. Hill, *Introduction to Statistical Thermodynamics*, (Addison-Wesley, Reading, Mass., 1960).
Terse but clear. Considers a more simple problem than that treated by Flory.
6.3   M. Volkenstein, *Configurational Statistics of Polymeric Chains*, (Translated from the Russian edn, S. N. Timasheff & M. J. Timasheff) (Interscience, New York, 1963).
Mathematical; considers further aspects of the problem.
6.4   A. G. Ogston, *Trans. Faraday Soc.*, **54**, 1754–1757, (1958).
A calculation of 'the space in a uniform random suspension of fibres'.

Chapter 7

Light scattering
7.1   H. V. Van de Hult, *Light Scattering by Small Particles*, (John Wiley & Son, New York, 1957, and also Chapman & Hall, London, 1957).
7.2   I. L. Fabelinskii, *Molecular Scattering of Light*, (Plenum Press, New York, 1968).
7.3   M. Kerker, *The Scattering of Light (and other Electromagnetic Radiation)*, (Academic Press, New York and London, 1969).

Laser scattering
7.4   B. Crosignani, P. di Porto, & M. Bertolotti, *Statistical Properties of Scattered Light*, (Academic Press, New York, San Francisco and London, 1975).

7.5   M. Weissman, H. Schindler & G. Feher, *Proc. Nat. Acad. Sci. USA*, **73**, 2776–2780, (1976).
Clear account of autocorrelation functions.

7.6   H. Z. Cummins & E. R. Pike, *Photon Correlation and Light Beating Spectroscopy*, (Plenum Press, New York and London, 1974).

Low-angle X-ray scattering
7.7   H. Brumberger (ed.), *Small Angle X-ray Scattering*, (Gordon & Breach, New York, London, Paris, 1967).

7.8   A. Guinier & G. Fournet, *Small Angle Scattering of X-rays*, (Wiley, New York, and also Chapman & Hall, London, 1955).

Neutron scattering
7.9   B. Jacrot, Study of Biological Structures by Neutron Scattering from Solution, *Rep. Prog. Phys.*, **39**, 911–953, (1976).

7.10 D. M. Engelman & P. B. Moore, Neutron-Scattering Studies of Ribosome, *Sci. Am.*, **235**, (4), 44–54, (1976).

Chapter 8

General
8.1   C. Tanford, *Physical Chemistry of Macromolecules*, (Wiley, New York, 1961).
Clearly written; deals with fundamentals.

8.2   H. Morawetz, *Macromolecules in Solution*, 2nd edn, (Wiley, New York, 1965).
More descriptive than Tanford's book.

Intrinsic viscosity
8.3   J. T. Yang, *Advances in Protein Chemistry*, **16**, 323–400, (1961).
A useful review of theory and methods.

Ultracentrifugation
8.4   T. Svedberg & K. O. Pederson, *The Ultracentrifuge*, (Oxford University Press, London, 1940).
A classical work, now somewhat dated.

8.5   H. K. Schachman, *Ultracentrifugation in Biochemistry*, (Academic Press, New York, 1959).
A standard work on practical and theoretical aspects.

8.6   T. J. Bowen, *An Introduction to Ultracentrifugation*, (Unwin, London, 1971).
A clear introduction emphasising practical aspects.

8.7   H. Fujita, *Mathematical Theory of Sedimentation Analysis*, (Academic Press, New York, 1962).
A lucid but mathematical approach to the theory.

8.8   J. W. Williams (ed.), *Ultracentrifugal Analysis*, (Academic Press, New York, 1963).
Useful articles on application of irreversible thermodynamics.

Diffusion
8.9   L. J. Gosting, *Advan. Protein Chem.*, **11**, 429 (1956).
A thorough review of many theoretical and practical aspects of diffusion.

Rotational diffusion
8.10 J. T. Edsall, *The Size, Shape and Hydration of Protein Molecules*, in H. Neurath & K. Bailey (ed), *The Proteins*, 1st edn, vol. IB (Academic Press, New York, 1953).
A good account of several methods.

Irreversible thermodynamics

8.11 I. Prigogine, *Introduction to Thermodynamics of Irreversible Processes*, (C. C. Thomas, Springfield, Ill., 1955).
An excellent, lucid introduction.

Hydrodynamics

8.12 H. Lamb, *Hydrodynamics*, 6th edn, (Cambridge University Press, Cambridge, 1932).
A standard work on an intrinsically mathematical subject.

Chapter 9

Thermodynamics of electrolytes

9.1 C. Tanford, *Physical Chemistry of Macromolecules*, (Wiley, New York, 1961).
Good accounts of Donnan equilibria and osmotic pressure.

9.2 R. A. Robinson & R. H. Stokes, *Electrolyte Solutions*, 2nd edn, (Butterworth, London, 1970).
A standard text. Chapter 2 is particularly relevant.

Polyelectrolytes

9.3 S. A. Rice & M. Nagasawa, *Polyelectrolyte Solutions*, (Academic Press, New York, 1961).
Mathematical and not for the beginner, but the chapter contributed by H. Morawetz is useful.

9.4 F. Oosawa, *Polyelectrolytes*, (Marcel Dekker, New York, 1971).
Lucid and non-mathematical.

Electrophoresis

9.5 M. Bier (ed), *Electrophoresis: Theory, Method and Applications*, (Academic Press, New York, 1959).

Counterion condensation

9.6 G. S. Manning, *Quart. Rev. Biophys.*, 11, 179–246, (1978).

Hydrodynamics and light scattering

9.7 H. Eisenberg, *Biological Macromolecules and Polyelectrolytes in Solution*, (Clarendon Press, Oxford, 1976).

Index

Page numbers in italics refer to sections in which the topic is discussed

259